D0307744

The History of Medicine

A Beginner's Guide

"As ever Mark Jackson offers us a humane and expansive view of the past to inform our vision of the future. As well as being a fantastic introduction to the history of medicine, this book is essential reading at a time when medical humanities scholars are for the first time working closely with clinicians and scientists to influence the direction of medical practice and therapy."

Jane Macnaughton – Professor of Medical Humanities, Durham University

"Clearly written, up-to-date and informative, it will allow any reader (as well as any student of the history of medicine) to better understand those ethical, moral and cultural questions we face daily in our interaction with health care in new and surprising ways."

Sander Gilman – Distinguished Professor of the Liberal Arts and Sciences, Emory University

"In one short book, Jackson has transformed the way we understand the theory and practice of medicine."

Joanna Bourke – Professor of History, Birkbeck College, University of London

"This lively and accessible book forms an exciting new resource, as it sets about instructing students on the importance of studying medical history and the factors which have determined and impacted on health and medicine over time."

Linda Bryder – Professor of History, University of Aukland

ONEWORLD BEGINNER'S GUIDES combine an original, inventive, and engaging approach with expert analysis on subjects ranging from art and history to religion and politics, and everything in-between. Innovative and affordable, books in the series are perfect for anyone curious about the way the world works and the big ideas of our time.

The History of Medicine

A Beginner's Guide

Mark Jackson

ONEWORLD

A Oneworld Paperback

Published in North America, Great Britain & Australia
by Oneworld Publications, 2014
Reprinted, 2016, 2020

Copyright © Mark Jackson 2014

The right of Mark Jackson to be identified as the Author
of this work has been asserted by him in accordance with
the Copyright, Designs and Patents Act 1988

All rights reserved
Copyright under Berne Convention
A CIP record for this title is available
from the British Library

ISBN 978-1-78074-520-6
eISBN 978-1-78074-527-5

Typeset by Siliconchips Services Ltd, UK
Printed and bound in Great Britian by Clays Ltd, Elcograf S.p.A.

Oneworld Publications
10 Bloomsbury Street
London WC1B 3SR
England

Stay up to date with the latest books,
special offers, and exclusive content from
Oneworld with our newsletter

Sign up on our website
oneworld-publications.com

For Ciara, Riordan and Conall

'A heart is what a heart can do.'

Sir James Mackenzie, 1910

Contents

List of illustrations

Preface

Writing this book has provided me with an opportunity to indulge two complementary aspects of my constitution: a passion for science and medicine on the one hand, and a commitment to history and the humanities on the other. Many years ago, in an earlier and more troubled life, I qualified in, and for a brief period practised, medicine. Although the problems of pathology and the challenges of clinical medicine held, and continue to hold, an enduring appeal, I found it increasingly difficult to cope at the time with the psychological and physical demands of the clinic and became frustrated by the narrow conceptual horizons of modern medical education, research and practice. After completing a doctoral dissertation on the history of legal medicine, a new pathway opened up for me as I began to explore, from a historical perspective, the social, cultural and political determinants of both medical knowledge and personal experiences of health and illness. This choice of career was fortuitous. A life in academia has allowed me to reconcile contradictory facets of my intellectual interests, to manage the vagaries of my personality as well as fluctuations in mood and energy, to engage constructively with colleagues and students, and to share more equally the pleasures and demands of marriage and parenthood with Siobhán.

My transition from dispirited physician to aspiring historian of medicine has been facilitated by numerous friends and colleagues. In particular, I have been encouraged and stimulated by the work of Roberta Bivins, Bill Bynum, Mike Depledge, Paul Dieppe, Chris Gill, Rhodri Hayward, Ludmilla Jordanova, Staffan Müller-Wille, John Pickstone, Roy Porter, Ed Ramsden, Matt Smith, Ed Watkins, John Wilkins and Allan Young, and by the continued support of postgraduate students, post-doctoral fellows and administrative staff in the Centre for Medical History, especially Fred Cooper, Barbara Douglas, Claire Keyte, John Ford, Jana Funke, Ali Haggett, Sarah Hayes, Grace Leggett, Kayleigh Nias, Debbie Palmer, Lyndy Pooley, Pam Richardson and Leah Songhurst. I am grateful to Steve Smith, Nick Talbot, Nick Kaye and Andrew Thorpe at the University of Exeter for their willingness to endorse my research initiatives. More directly, the form and content of this book have been strongly informed by lectures and seminars delivered to undergraduate and postgraduate students at Manchester and Exeter, and by various opportunities to teach GCSE and A level students at Colyton Grammar School and other local schools and colleges. I am grateful to all those students for listening, arguing, and sometimes agreeing.

Of course, none of these research and teaching activities would be possible without the enlightened support of the Wellcome Trust, which has been instrumental in shaping not only the emergence of the history of medicine in recent decades but also the growth of medical humanities. The Trust's commitment to rendering historical research instrumental in the pursuit of better health has generated opportunities for greater, and more meaningful, exchange between science and the humanities and helped to sustain my dream of uniting disparate aspects of the practice of medicine. I am also grateful to Wellcome Images for providing, and allowing me to reproduce, the illustrations and to Caroline Morley and Miriam Ward for their assistance. I am greatly indebted to Fiona Slater of Oneworld, who first suggested that I write a book of this nature on this topic. Fiona's vision and advice have been crucial both to the initial formulation and the eventual realization of the work. I would also like to thank Ann Grand for her meticulous copy-editing and David Inglesfield and Linda Smith for their careful proof-reading.

Although Siobhán will forever be my muse in all spheres of life, it is our children who especially challenge my inherent tendency to solipsism and misanthropy. While they undoubtedly share my genetic constitution, they display their own idiosyncratic intelligence, temperament and sensitivity in a way that forces me to view the world with fresh and, I hope, more considerate eyes. This book is written for Ciara, Riordan and Conall, who remind me regularly of the heart's capacity to adapt to new demands.

Introduction

'By the historical method alone can many problems in medicine be approached profitably.'

William Osler, *Aequanimitas*, 1928

Medicine touches us all at some stage in our lives. Whether we live in a crowded high-tech westernized society that uses the diagnostic and therapeutic tools of modern bioscience or in an isolated rural community where health care is perhaps less formal, less intrusive and less commercial, it is arguably medicine, rather than religion or law, that dictates the manner in which we are born, the quality of our lives, and the ease and speed of our deaths. Indeed, although modern populations are increasingly struggling to cope with chronic conditions such as cancer, heart disease, arthritis, obesity and depression, we have come to rely heavily on the ability of medicine to help us live relatively happily, healthily and productively well into our eighties.

Given the extent to which it penetrates the physical, psychological and even spiritual dimensions of human existence, it is no surprise that medicine constitutes a vast territory. In the early twenty-first century, the practice of medicine incorporates, among other things, the preservation of health and the prevention of illness, the discovery and application of pharmacological tools to combat mental and physical disease, the development of novel diagnostic and surgical techniques to identify and remove tumours, heal broken bones or restore blood-flow to ailing hearts, the formulation of policies designed to protect national and global public health, the use of psychotherapy to reduce depression and anxiety and to promote happiness, the delivery of welfare services and medical support to mothers and their children, and the alleviation of pain and disability.

In the past, the contours of medicine have been even more expansive. In both Eastern and Western cultures, medicine has embraced religion, magic, alchemy and astrology, as well as the application of herbal remedies, the use of healing rituals, sacrifices and offerings to the gods, and the relief of poverty. Health care has been dispensed not only in specialist institutions, including hospitals, workhouses, monasteries and hospices, but also regularly in the community, on the battlefield and at home. Within these diverse environments, advice and treatment were delivered by a range of practitioners often trained in quite different ways and possessing different, although usually complementary, skills and knowledge. In sickness and in health, patients sought the services of shamans, diviners, priests, midwives, nurses, physicians, surgeons, apothecaries, and a miscellany of itinerant practitioners, charlatans and quacks. Historically, medicine has never constituted a monolithic system of knowledge and practice but has always been marked by a vibrant sense of diversity and pluralism.

The task of unravelling the history of medicine is further complicated by the fact that medical theory and practice, as well as the distribution and patterns of disease, have been so deeply embedded in social contexts that the boundary between medicine and society has been indiscernible. In all ages and all cultures, the appearance, spread and control of both infectious and non-infectious diseases have been dictated by social, economic and cultural factors. At the same time, the practice of medicine has been a social endeavour, not only reflecting the norms and expectations of patients and politicians alike but also influencing the beliefs, customs and hopes of the sick, the healthy and their healers. Even in the modern era of biomedicine, when science appears to offer a more objective perspective on health and illness, scientific knowledge, clinical practice and health-care policies continue to be determined by social and cultural factors as well as economic and political expediency.

There has been a tendency in recent times to distinguish rather deliberately between science and the humanities, as if they possess entirely different agendas and methods or constitute entirely different intellectual cultures. While science and medicine appear to offer more reliable accounts of the natural world and its problems, the humanities seem to deal only with subjective, and often unverifiable, aspects of personal and public life. As a result, historical, philosophical or literary studies of medicine and science have often been divorced from the pursuit of clinical knowledge, improved health policies and better treatments. For a number of reasons it is a mistake to impose a distinction between medicine and history in this way. In the first instance, the notion of history has always been integral to clinical method. From ancient to modern medicine, students have been taught to consider the patient's history from various perspectives: the history of current symptoms; the patient's past medical, occupational and social history; and the family (and increasingly this means genetic) history. Personal and biological, as well as collective and psychosocial, histories have thus been central to the processes of accurately diagnosing disease and formulating appropriate treatments and policies. Second, as both historians and doctors have pointed out, history also constitutes a vehicle for educating, inspiring and humanizing medical and nursing students who might otherwise succumb to the brutalizing effects of regular exposure to disease and death.

Perhaps more contentiously, research in the medical humanities allows us to recognize the power and limits of medicine and to acknowledge the cultural, social and political, rather than merely technical, obstacles to health promotion and disease prevention. By exploring the human aspects of medicine and tracing the development of medical theories, policies and institutions across time, medical history can reveal the manner in which medicine reflects and shapes far wider historical currents and the extent to which experiences of health and disease structure our lives. More broadly, while science can uncover many of the mechanisms underlying patterns of health and disease, it is the humanities that can more effectively reveal the meanings of our experiences of pain and suffering. Medical history and the wider humanities, like the biomedical sciences, should therefore be integral to our search for health and happiness.

Historians have approached the history of medicine in different ways. Some scholars have focused on narrating and celebrating great discoveries made by pioneers in the field or on the health and illnesses of key historical figures. In these stories of progressive innovation, the achievements of Hippocrates, Galen, Ibn Sīnā, Ambroise Paré, Andreas Vesalius, William Harvey, Edward Jenner, John Snow, Ignaz Semmelweis, Florence Nightingale, Joseph Lister, Louis Pasteur, Robert Koch, Alexander Fleming and many others have taken centre stage. Such tales of success are not without merit: they highlight the extraordinary contributions of doctors to the history of humankind and bring the drama and significance of medicine to the fore. At the same time, however, they often give precedence to the accomplishments of men over women, the traditions of the West over the East, and the importance of biological and technological, rather than social and cultural, factors.

By contrast, social historians have recently moved away from telling stories of triumphal progress towards an approach that emphasizes the historical contingency of medical knowledge and the cultural specificity of experiences of health and illness. In these histories, there is no fundamental or enduring truth waiting to be unearthed by enlightened scientists and doctors; rather, knowledge and practice are regarded as always shifting, and contested, products of socio-cultural and political forces. While such accounts of medicine and disease in the past effectively reveal the social determinants of health and healing, they tend to lose the sense of theatre and urgency embedded in the practice of medicine and to ignore the extent to which both past and present populations have routinely depended on medicine to forge a better world.

In many ways, this book is an attempt to establish a middle way between these distinctive, and sometimes competing, formulations of medical history. Given that the past is unknowable, or at least inaccessible to direct perception, any account of the history of medicine is dependent on the precise perspective from which it is written and on the selection and analysis of particular sources. Coloured by my training as a doctor and my life as an historian, the following chapters are predicated on a belief that medical history and biomedical science are complementary tools in our attempts to engineer significant improvements in the health of human populations. History not only reveals elements of continuity and change in medical theory and practice but also exposes the close relationship between personal experiences of illness, scientific knowledge of bodies and minds, and the broader social factors that influence our understandings of health and disease. In addition, historical research clearly demonstrates shifting attitudes to the complex interactions among patients, doctors and disease. Given the vast geographical and chronological range encompassed by the term 'medical history', the narrative is necessarily determined by my idiosyncratic intellectual and clinical interests. Nevertheless, I have attempted to incorporate Eastern, as well as Western, medical traditions, to trace the distinctions between alternative and orthodox practices, and to explore the similarities and contradictions between lay and expert knowledge. My aim is to reveal the manner in which the boundaries between health and disease, between science and the humanities, and between the past and the present are less secure than we often imagine.

1

Balance and flow: the ancient world

The woman who lodged at the house of Tisamenas had a troublesome attack of iliac passion, much vomiting; could not keep her drink; pains about the hypochondria, and pains also in the lower part of the belly; constant tormina; not thirsty; became hot; extremities cold throughout, with nausea and insomnolency; urine scanty and thin; dejections undigested, thin, scanty. Nothing could do her any good. She died.

Hippocrates, *Of the Epidemics*

It is the fifth century BCE. A young man from the coastal city of Abdera is being examined by Hippocrates, the most famous physician in Greece. For several days, the patient has been suffering from an acute fever accompanied by pain in his right side, a dry cough, thirst and increasing delirium. Having failed to relieve the pain and cough with warm applications, on the eighth day Hippocrates opens a vein at the elbow to release blood. This treatment seemed to promote recovery. According to Hippocrates' subsequent report of the case in his extensive work on epidemics, the patient's pains, fever and thirst diminished, he began to cough up sputum, and his breathing eased. Unlike many less fortunate patients in the ancient world, by the thirty-fourth day of his illness the young man was fully restored to health.

Historians have often regarded the Greek physician Hippocrates (c. 460–371 BCE) as the father of medicine. There is some truth to this claim. Hippocrates and his students laid the foundations of an approach to medicine in the Western world that lasted over 2,000 years. As the case of the young man from Abdera demonstrates, the works attributed to Hippocrates provided early descriptions of many diseases and injuries, including epilepsy, epidemics, ulcers and fractures, and established a range of preventative and therapeutic interventions, including herbal remedies, surgery, bleeding and dietary recommendations. In addition, they explored the impact of environmental conditions on health and provided a framework for professional ethics encapsulated in the Hippocratic Oath, which in many countries remains a key feature of the process of induction into the medical profession.

The pivotal place of Hippocrates at the origins of medicine can, however, be challenged. Many of the books attributed to Hippocrates, referred to as the Hippocratic Corpus, were not written by him. In addition, there was no single Hippocratic system even in the ancient West: Hippocratic theories were contested and shifted over time in much the same way that medicine has evolved and been challenged in the modern world. Finally, and perhaps most importantly, there were earlier formulations of diseases and their treatments in the ancient Eastern cultures of China, India and Egypt that arguably had a longer-lasting influence on the practice of medicine than those of Hippocrates. Indeed, these overlapping Eastern traditions may well have informed Hippocratic medicine, which can be regarded in some ways as derivative rather than entirely original.

It is difficult to reconstruct ancient medical theories and practice with any certainty. Textual sources are often fragmentary, difficult to decipher or available only in translation, and bio-archaeological evidence is open to interpretation. In spite of these challenges, the sources suggest that, although there were differences in the manner in which ancient cultures understood, experienced and treated disease, there were also common elements in Chinese, Indian, Egyptian and Graeco-Roman medicine. Most ancient medical systems around the world explained disease in terms of the impact of constitution, lifestyle and environment on the balance and flow of humours or energy through designated channels in the body. Ancient approaches to the prevention and treatment of ill-health were also similar, focusing on the application of herbs, the restoration of humoral balance, the occasional use of surgery, and appeals to the benevolence of supernatural forces. To understand the emergence and development of these diverse, but overlapping, medical traditions we need to consider three broad questions. Which diseases did ancient populations live with and die from? How did patients and their doctors or healers explain and experience health and illness? And how did ancient doctors in the East and West attempt to prevent and treat disease?

Yin and yang

The origins of Chinese medicine are shrouded in myth. The key ancient Chinese medical texts are attributed to two legendary sages from the third millennium BCE, Huangdi, the Yellow Emperor, and Shennong, also known as the Divine Farmer or Red Emperor. Based on oral traditions, the works of these ruling sages were not formalized in written texts until the Han dynasty, between 206 BCE and 220 CE. While Huangdi's *Classic of Internal Medicine* (or *Neijing*), written in the form of a conversation between the Emperor and his physician, constituted an account of the principal causes and symptoms of disease, Shennong's *Materia Medica* (or *Pen-ts'ao ching*) provided an extensive treatise on the classification and therapeutic use of food and drugs.

These canonical texts suggest that the ancient Chinese suffered from a similar range of acute and chronic diseases to many modern populations, although they were usually described broadly in terms of symptoms or altered function, rather than in terms of separate disease categories. Epidemics of fevers or infectious diseases were common, as were diseases of the heart, lungs, eyes, stomach and liver, and illnesses attributed to emotional or psychological disturbances such as insanity, anger, numbness and loss of speech. For example, the condition of laboured breathing, which became known in the West as asthma, was recognized in early Chinese texts. However, in contrast to modern Western medicine, such conditions were regarded as the product of systemic alterations throughout the body, not merely as the result of damage to specific organs.

Chinese medical theories of health and disease focused on the flow of Qi (usually translated as force or energy) around the body through designated channels (*mai*) and on the balance between the complementary properties of *yin* and *yang*. Well-being was a state in which the flow of Qi was uninterrupted and there was an appropriate state of equilibrium between *yin* and *yang*. Disease was the result of disturbances in the flow of Qi and imbalances between the quality and quantity of *yin* and *yang*, generated for example by over-indulgence. The properties of *yin* and *yang* did not correspond with particular components of the body but, rather, represented alternative aspects of bodies and the universe: *yin*, the female element, signified darkness, dampness, cold, the moon and death; by contrast, the male element, *yang*, symbolized the sun, heaven, light, heat and life. An appropriate mixture of *yin* and *yang* was essential for human health, political stability and the harmony of the cosmos. Within this comprehensive scheme for explaining life and death, which owed much to the ancient philosophical concept of *tao* ('the way'), the body not only constituted a microcosm of creation but was also considered analogous to the organization of the empire: bodies, like countries, were run by rulers (comparable to the 'heart-mind'), state officials (the lungs) and others (the liver, kidneys, spleen and intestine) working together to ensure peace and stability.

Ancient Chinese medicine comprised an amalgam of anatomical and physiological understandings of the body and a belief in both harmful and healing spirits. Chinese doctors included shamans and diviners (known as *wu*), priests (or *chu*), practitioners of demonic medicine and a variety of less respected itinerant practitioners and female healers. The role of healers was to diagnose disease, predict the outcome of disorders and provide effective treatment. Diagnosis was based primarily on examining the pulse at various locations on the surface of the body, as well as observing skin colour, emotional state, the pattern of breathing, and the condition of the tongue and sensory organs. The quality of the pulse, which remains a major concern of traditional Chinese medicine in the modern world, was particularly important, since it revealed the flow of *Qi* through the body. Accurate prognosis of the outcome of disease depended not only on close examination of the body but also on recognizing the impact of environmental conditions on the appearance and symptoms of disease, on medical divination and numerology, and on understanding the relationship between the state of the body and the cycles of nature.

Treatment involved restoring the flow of *Qi* and re-establishing the balance of *yin* and *yang* to release energy. A variety of

---

HEALTHY PULSE, HEALTHY HEART

According to the Yellow Emperor's physician, Ch'i Po, health and sickness could be determined by feeling the pulse:

'When man is serene and healthy the pulse of the heart flows and connects, just as pearls are joined together or like a string of red jade – then one can speak of a healthy heart.

'When man is sick the pulse of his heart rushes and pants. When this panting is continuous and springs from within and the pulse beats are wrong and small – then one can speak of a sick heart.'

Huangdi's *Classic of Internal Medicine*, third millennium BCE

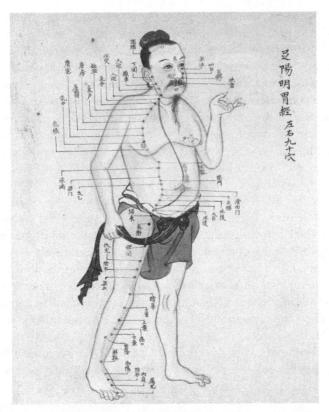

Figure 1   Chinese acupuncture chart (Wellcome Library, London)

approaches was adopted by ancient Chinese healers. The most important technique was acupuncture, which involved the insertion of fine needles into designated points on the body to encourage the unimpeded flow of *Qi*. Most medical texts contained diagrams depicting the appropriate acupuncture points used to treat particular conditions (see Figure 1). An alternative to acupuncture was moxibustion, in which preparations of *moxa* (mugwort) were burnt at key points on the skin to unblock the obstructed flow of *Qi*, particularly in chronic conditions. Such approaches to preserving or restoring the flow of blood and *Qi* through the bodily channels remained pivotal not only to the practice of modern Chinese medicine in the East but also to the subsequent adoption of Chinese techniques in many Western countries.

Classical Chinese medical texts reveal other forms of treatment. Although neither surgery nor anatomical dissection was used, except for the occasional treatment of injuries and abscesses, blood-letting was employed to restore balance and flow. Drugs derived from animals, vegetables and minerals, taken in pills, powders and syrups, also constituted important tools in the attempt to prevent and cure disease. Therapeutic minerals, which were produced and administered according to alchemical principles and valued according to their rarity, included gold, lead, mercury, arsenic and cinnabar. Herbs were classified according to their colour, aroma and taste, and usually mixed to produce therapeutic 'decoctions'. As in many other ancient cultures, plant and animal substances were sometimes prescribed according to the principle of correspondence; that is, according to the belief that the healing properties of a substance were related to its resemblance to an organ or attribute of the body or to the symptoms of a specific disease. Tiger bone or rhinoceros horns, for example, were thought to impart strength. Some traditional Chinese medicines form the basis of modern drugs: *ma huang* (ephedra) was prescribed by ancient Chinese doctors to relieve breathing difficulties; in the form of ephedrine this is still used in the twenty-first century to relieve hay fever and acute attacks of asthma.

Existing alongside these pharmacological approaches were more gentle treatments such as massage, exercise and dietary supplements. In addition, many practitioners and their patients endorsed spiritual approaches to health and illness, informed initially by Daoism and later by Buddhism. The practice of *dao yin*, which was perhaps similar to yoga or gymnastics, used a combination of breathing exercises, physical movement and introspection to purify the body and restore vitality. Religion provided a sense of continuity and structure to the practice of Chinese medicine, in spite of the fact that doctors and their patients were increasingly adopting more secular understandings of health and disease. Ritual, worship, confession, meditation and sacrificial offerings were routinely used, particularly by the lower social classes, to prevent and treat diseases thought to be caused either by sin or by demons and evil spirits.

It is important to recognize that ancient Chinese medicine was neither singular nor static, as some historians and practitioners have assumed. Although the works of Huangdi and Shen-nung remained central to the practice of medicine, particularly amongst elite physicians, their descriptions and prescriptions were adapted by later writers such as Zhang Zhongjing (second to third century CE), whose account of fevers was for many centuries employed by governments attempting to prevent and combat fatal epidemics. In some parts of China, it was the Tibetan 'science of healing' (*sowa rigpa*), with its mix of humoral theory and a belief in spirits and the protective value of mantra, exorcism and ritual, that became the most common form of medicine. Equally, different forms of medicine were practised across social classes: while the upper classes employed learned physicians who were conversant with classical texts, members of the lower social classes consulted a wider range of practitioners, who offered cheaper alternatives to élite medicine. However, in all forms of medical practice, etiquette was important. Doctors were often prohibited from touching patients of the opposite sex, except to detect the strength, rhythm and character of the pulse.

The influence of traditional Chinese medicine was not confined to China. By the sixth or seventh century CE, the exchange of texts and people across regional borders had facilitated the spread of Chinese approaches to health and disease, along with Buddhism, to Korea and Japan, partially replacing the Japanese reliance on ritual medicine performed by Shintō healers. Both Korean medicine (*hanui*) and Japanese medicine (*kanpō*) adopted the holism associated with Chinese medical practice: infectious diseases, such as plague, smallpox, measles and influenza were regarded as the result of functional imbalances in the body and interpreted in the light of beliefs that all living beings operated in accordance with the rules of nature and the cosmos. Korean and Japanese modes of treatment were also based on a mixture of secular and spiritual strategies. Like their Chinese predecessors and neighbours, doctors in ancient Korea and Japan applied acupuncture, moxibustion, herbs, massage and gymnastics to maintain health and treat disease. In this way, the ancient theories and practices first propounded by the Red and Yellow Emperors provided the basis for medicine throughout medieval China, Korea and Japan.

Living healthily

Ancient Indian and Sri Lankan forms of medicine were based on a healing tradition that had its origins in the third or second millennium BCE. Indigenous medical knowledge was based on Sanskrit liturgical documents, the contents of which had been imparted by the Hindu god Brahma and translated by the Vedic sages into a system known as āyurvedic medicine. The principal texts were the *Caraka Saṃhitā* and the *Súsruta Saṃhitā*, both of which were probably compiled during the first millennium CE and together formed the cornerstone of āyurvedic practice. These seminal āyurvedic works were translated into Tibetan, Arabic and English, facilitating the spread of ancient Indian medical wisdom to other regions of the world. Meaning literally 'knowledge for long life', āyurveda constituted not only a form of medical care but also a way of living healthily and happily. While āyurveda flourished in the northern Indian provinces, an alternative form of esoteric magical and alchemical healing, known as Siddha, emerged in the Tamil-speaking areas of southern India.

Our understanding of ancient āyurvedic practices comes not only from medical texts but also from archaeological evidence. These suggest that ancient Indian populations suffered from a range of diseases and injuries, including epidemics of infectious diseases, skin disorders, dropsy, epilepsy, asthma, consumption and insanity. The causes and mechanisms of disease were explained in terms of humoral and spiritual balance: the body was thought to be composed of three humours (*dosas*), namely wind, bile and phlegm, and seven bodily constituents: blood, chyle, flesh, fat, bone, marrow and semen. Any disruption in the quality and quantity of the humours and elements resulted in disease. In addition, illness could be caused by evil spirits. Within this medico-spiritual scheme, lifespan was thought to be determined not only by current behaviour, including poor diet, over-exertion, excessive sexual activity and abnormal posture, but also by past deeds.

Accurate diagnosis depended initially on physical examination and analysis of the pulse. By the eleventh century CE, these procedures were supplemented by urine analysis. Prevention and treatment of disease depended on the patient maintaining or regaining appropriate measure and balance. This could be achieved by pursuing a life of moderation, rather than excess, with regard to sleep, exercise, diet and sex. When illness intervened, to restore health, healers prescribed a range of herbal preparations, minerals, vegetables and animal substances, particularly dung from cows, which were holy animals to Hindus. Hygiene was also considered important for the maintenance of health: there is archaeological evidence that some ancient Indian populations used communal baths, cleansing rites and drainage systems to prevent the spread of disease. As the *Súsruta Samhitā* makes clear, in addition to blood-letting, āyurvedic practitioners also employed fairly advanced surgical techniques, particularly the removal of cataracts and kidney stones, the treatment of injuries sustained in battle, and suturing. Proficiency in surgery and knowledge of anatomy were obtained partly from the examination of cadavers.

Ancient Indian medical texts also set out the qualities and behaviour expected of good practitioners. The virtuous healer was required to be knowledgeable, courteous and self-disciplined – characteristics that would not only be appreciated by patients but also ensure the satisfaction and happiness of the doctor (*vaidya*). During an oath of initiation, āyurvedic students swore to remain celibate, to speak truthfully and maintain confidentiality, to follow a vegetarian diet, and to treat women only in the company of a husband or guardian. Unscrupulous practitioners or quacks, those who tended to insult their medical colleagues and ingratiate themselves with patients for their own benefit, were to be condemned.

Vedic medicine was not the only system of preventative and prescriptive health care in ancient India. Patients seeking treatment could choose between orthodox āyurvedic medicine; alternative forms or variants of āyurvedic practice; shamanistic approaches to illness, based largely on folk traditions; and various forms of astrological medicine, which regarded children's diseases in particular as the result of malign forces or demons. In addition, while the influence of āyurvedic medicine clearly persisted, over the centuries it was subject to challenge and alterations from competing formulations of health and illness that served to identify new diseases, develop novel diagnostic techniques and establish fresh approaches to treatment. Like traditional Chinese medicine, ancient Indian medicine was therefore neither uniform nor unchanging. Indeed, in the eleventh century CE, indigenous Indian systems of healing were transformed by exposure to the principles of Greek and Islamic medicine, which were imported during the Afghan invasions of India and which formed the basis of ūnānī tibb, an Indian term derived from Greek 'Ionian' medicine.

Plants and papyri

Historians and archaeologists have used evidence from palaeo-pathology and the study of papyri to reveal patterns of disease and the cardinal features of Egyptian medicine. Archaeological studies of mummies, skeletons and clay statuettes excavated from various sites in Egypt suggest that ancient Egyptian populations suffered from a variety of conditions: intestinal infections, lung diseases such as tuberculosis and pneumoconiosis, hardening of the arteries, tumours, obesity, arthritis, wounds, dental disease, mental illness and various genetic conditions such as dwarfism have all been demonstrated. In addition, the presence of trans-verse 'Harris lines' on the long bones of children testifies to the arrested growth caused by malnutrition and the stress of natural disasters. Egyptian women usually married at the age of twelve or thirteen and often suffered from birth injuries, such as fistulas and prolapses, which sometimes proved fatal.

Further evidence from the papyri, which were composed either in hieroglyphs or in the hieratic script, reinforces these archaeological findings. Most of the papyri were discovered in the late nineteenth century during expeditions funded and led by Western archaeologists and anthropologists and were named in different ways: in recognition of the geographical region in which they were found, after the archaeologists and dealers who purchased them, or according to the museum or library where they were eventually deposited. The oldest of the Egyptian medical treatises is the Kahun papyrus, which dates from approximately 1900 BCE and focuses on the diseases of women and contraception. The Edwin Smith papyrus, dating from 1600

BCE and named after the dealer who bought it in 1862, sets out the features and treatment of surgical cases, while the Ebers papyrus, compiled in about 1550 BCE and purchased by the German Egyptologist Georg Ebers (1837–98) in the 1870s, is a compendium of treatments for various diseases. Other papyri, including the Berlin, Chester Beatty, Hearst and London papyri, provide a mixture of medical and magical remedies, including incantations and spells. Although magic and religion played a part in their understanding of disease, Egyptian doctors (*sinw*) and their patients, like their counterparts in China, India and Greece, explained illness predominantly in terms of humoral imbalance. The accumulation of phlegm (*stt*) in the lungs, for example, led to respiratory diseases such as bronchitis, asthma and colds.

Our knowledge of the practice of Egyptian medicine is largely limited to the work of elite physicians to the pharaohs, such as the chief physician and dentist Imhotep and the court physician, Iry. Doctors often trained and administered health care in healing temples, where patients would not only receive medicine but were also expected to offer sacrifices to specific gods or wear amulets to ward off sickness caused by evil spirits. Each speciality within medicine had a dedicated deity, such as the goddesses of childbirth, Hathor and Taurt. These deities not only determined the appearance and severity of disease but also directed the treatments prescribed by doctors. One of these gods, Horus, continues to influence modern medicine. The Eye of Horus (*Wadjet*) was believed to have protective and therapeutic properties and a variant of the hieroglyph is still used as shorthand by Western doctors to signify a prescription or treatment.

Egyptians living on the banks of the Nile, in large cities or employed in the working parties constructing pyramids and state buildings, were prone to famine and flooding, with the inevitable accompanying infestation of rats and the spread of infectious diseases. Egyptian approaches to health and illness included measures designed both to prevent and cure disease. Hygiene practices aimed at reducing the impact of epidemics amongst crowded populations included the appointment of doctors to oversee workmen, the provision of sufficient food and the regular use of soap to improve personal appearance and cleanliness. In a ritual that pre-dated modern cosmetic fashion by over two millennia, the last of the Egyptian pharaohs, Cleopatra, is thought to have covered herself in warm wax to remove unwanted body hair. Ordinary Egyptian men and women used razors to shave and tweezers to pluck hairs, as well as applying cosmetics such as kohl and malachite. Kohl is a dark pigment derived from lead sulphide, initially used to protect the eyes from sun damage but subsequently employed, particularly by Egyptian queens, as an eyeliner or eyeshadow.

More specific treatments were recommended for particular conditions. There is some evidence from excavated skulls that dental abscesses caused by dental attrition were treated by drilling boreholes in the jaw. It is likely that surgeons regularly accompanied Egyptian armies, offering medical advice and dressing wounds. In addition, the Edwin Smith papyrus gives precise instructions on how to treat a dislocated jaw, and how to stem the blood flow and reduce the swelling of a broken nose. But it was plants that dominated the management of many diseases. The papyri suggest that Egyptian doctors advised their patients to ingest or inhale a wide range of herbal remedies: mandrake for its pain-relieving and aphrodisiac properties, resin from fir trees to clean infected wounds, aloe to remove phlegm and relieve catarrh, cinnamon to soothe ulcerated gums, and henna to delay or reverse hair loss. Honey and beer also became pivotal ingredients of Egyptian medicines, partly as vehicles for the administration of drugs: the roots of the marshmallow plant, which grew on the river banks and in salt marshes in Europe and western Asia, were combined with honey to produce a popular remedy for sore throats. Some of these Egyptian drugs were adopted by other healing traditions, such as ancient Greek and Roman or medieval medicine, and they continue to be used in some form by both orthodox and alternative practitioners in the modern world. The evidence indicates that Egyptian medicine, rather than being seen as primitive, should be regarded as remarkably sophisticated.

Perhaps the most notable feature of ancient Egyptian civilization, apart from the pyramids at Giza, is the Egyptian attitude to the body after death. Strong belief in an afterlife led the Egyptians to make extensive preparations for the journey of the body across the threshold from life to death. Major organs, excluding the heart, because it was thought to be the seat of the soul, were removed and placed in Coptic (or Canopic) jars, which were often engraved with the four sons of Horus, and buried with the mummified bodies. This process suggests that the Egyptians had a relatively advanced knowledge of anatomy. Even when improved embalming techniques made removal of the organs unnecessary, Coptic jars continued to be placed near the sarcophagus, ensuring the persistence of an Egyptian medico-religious tradition that had started approximately 2,000 years previously.

Blood, phlegm and bile

Born on the island of Kos into a medical family, Hippocrates dominated ancient Greek and Roman medicine. Surviving sources from ancient Greece, in particular the works of Galen (c. 129–210 CE) and Dioscorides (c. 40–90 CE) as well as Hippocrates, are fragmentary but, together with archaeological and palaeopathological evidence, testify not only to the complexity of Greek theories of medicine and classifications of disease but also to the range of diseases suffered by ancient Greek populations, the variety of approaches to prevention and treatment, and the ethics of good medical practice. As in earlier and parallel ancient medical traditions, Graeco-Roman medicine included elements of rational secular and natural understandings of the body and cosmos on the one hand, and practices that appear to modern sensibilities to be irrational magico-religious beliefs on the other.

Although Hippocrates became the archetypal ancient doctor and was referred to by Plato (428–348 BCE) as 'the famous physician', it is important to recognize that even before Hippocrates many Greek doctors and natural philosophers had provided a theoretical and practical basis for Graeco-Roman medicine. Influenced partly by features of Egyptian and Babylonian medicine that had been imported by itinerant doctors, physicians such as Democedes (sixth century BCE), Alcmaeon of Croton (490–430 BCE) and Apollonides (fifth century BCE) established the importance of humoral balance or symmetry within the body and identified many natural herbal remedies for specific symptoms. Nevertheless, it is true that most of our knowledge of Greek medicine comes from the Hippocratic Corpus, much of which was subsequently translated, analysed and interpreted by other ancient (and eventually medieval) authors such as Erasistratus (c. 315–240 BCE), Celsus (25 BCE–50 CE), Aretaeus of Cappadocia (c. first to second centuries CE) and, most importantly, Galen.

The Corpus comprises nearly sixty treatises on the causes, prevention and treatment of sickness. Some of the works focus on specific types of injury and disease, including epidemics, epilepsy, fractures and diseases of women. Others either offer advice on particular clinical problems, including the difficulties of accurate prognosis and the importance of regimen or lifestyle in determining the symptoms and outcome of disease, or provide philosophical but pragmatic reflections on contemporary theories of disease, including close attention to the relationship between health and the environment. One of the most influential elements of the Hippocratic Corpus is the collection of memorable aphorisms that became a standard educational text for medieval and Renaissance medical students. In addition to explaining the difficulties of practising medicine, the aphorisms provided accessible advice on many aspects of diagnosis, prognosis and treatment: 'When a

THE HIPPOCRATIC OATH

I swear by Apollo the physician, Asclepius, Hygeia, and Panacea and I take to witness all the gods, all the goddesses, to keep according to my ability and my judgment, the following Oath and agreement:

To consider dear to me, as my parents, him who taught me this art; to live in common with him and, if necessary, to share my goods with him; to look upon his children as my own brothers, to teach them this art; and that by my teaching, I will impart a knowledge of this art to my own sons, and to my teacher's sons and to disciples bound by an indenture and oath according to the rules of the profession, and no others. I will prescribe regimens for the good of my patients according to my ability and my judgment and never do harm to anyone. I will give no deadly medicine to anyone if asked, nor suggest any such counsel; and similarly I will not give a woman a pessary to cause an abortion. But I will preserve the purity of my life and my art. I will not cut for stone, even for patients in whom the disease is manifest; I will leave this operation to be performed by practitioners, specialists in this art. In every house where I come I will enter only for the good of my patients, keeping myself far from all intentional ill-doing and all seduction and especially from the pleasures of love with women or men, be they free or slaves. All that may come to my knowledge in the exercise of my profession or in daily commerce with men, which ought not to be spread abroad, I will keep secret and will never reveal. If I keep this oath faithfully, may I enjoy my life and practise my art, respected by all humanity and in all times; but if I swerve from it or violate it, may the reverse be my life.

person who is recovering from a disease has a good appetite but his body does not improve in condition', warned Hippocrates, 'it is a bad symptom'. Perhaps the most well-known element of the Corpus, however, was the Hippocratic Oath, which set out the standards of behaviour expected of medical practitioners and has retained its legal and ethical significance into the modern period.

In Hippocratic medicine, health and disease, as well as character and personality, were thought to be determined not by divine interference but largely by balances or imbalances in the four bodily humours: blood, phlegm, yellow bile and black bile. An excess or depletion of one or more of the humours resulted in obstruction or dysfunction of the organs, leading to the signs and symptoms of disease. Although many of his peers regarded epilepsy as a sacred or divine disease, for example, Hippocrates insisted that its cause lay simply in the accumulation of phlegm in the brain. It is important to note that the natural system of balance embedded in Hippocratic formulations of humoralism was not confined to medicine. According to many ancient authors, the four humours were correlated to the four elements (earth, air, fire and water), the four primary qualities (hot, dry, cold and wet), the four temperaments (phlegmatic, melancholic, sanguine and choleric), and the four seasons. As in Chinese and āyurvedic medicine, balance and symmetry were key Greek concepts in explaining the capacities of individuals, societies and the cosmos, as well as health and disease.

The doctrine of the humours was employed to explain not only individual, and often hereditary, predispositions to certain diseases but also the susceptibility of particular groups of patients to particular diseases. Women and children were considered to be cooler and moister than men and therefore more vulnerable to diseases caused by an excess of phlegm, such as dysentery, asthma and catarrh. Humoral theory also lay at the heart of an approach to health care that focused on environmental conditions. Patterns of health and illness in different geographical regions were thought to be determined by the strength and direction of the wind, by the location of clean and polluted water sources, and by seasonal changes in the weather. In his treatise on *Airs, Waters and Places*, Hippocrates offered advice to people of different constitutions on where to live at certain times of the year to alleviate the symptoms of particular diseases, thereby establishing what could be regarded as an early form of medical meteorology.

The Hippocratic Corpus contained classic descriptions of the many diseases that afflicted ancient Greek populations, including consumption, tetanus, pleurisy, asthma, diabetes, certain types of cancer and various forms of mental illness such as mania and melancholy, which were similarly attributed not to divine intervention but to the effects of bile and phlegm in the brain. Ancient Greek populations were decimated by wars, famine and plagues, which were sometimes spread by new agricultural practices and domestic animals or brought in along trade routes. Indeed, the growing economic reliance on livestock, including sheep and goats, may well have encouraged the spread of bacterial and parasitic diseases, such as anthrax, plague and rabies. The ancient Greeks were also more likely to suffer from illnesses caused or exacerbated by poor sanitation, inadequate diets, unhealthy occupational conditions and, on occasion, profligate lifestyles. The prevalence of disease and high infant mortality ensured that the average life expectancy of Greek and Roman populations was in the region of only twenty or thirty years. However, if someone survived beyond infancy they might live to the age of sixty.

Ancient Greek and Roman patients consulted a wide range of healers, including physicians, midwives, medical attendants or physical trainers, herbalists and astrologers, who offered advice not just on matters of health and sickness but also on the maintenance of vitality, fitness and appearance. For the most part, élite physicians were men but there were some women practitioners who provided treatment to other women, particularly in the management of childbirth. In some locations the fees charged by doctors and midwives could be regulated by the state. Although the social and professional boundaries between the various categories of medical practitioners were not always clear-cut, male physicians were assumed to possess superior moral character, higher educational qualifications and more extensive professional experience, and they certainly enjoyed greater official recognition from civic authorities than other practitioners. In this way, the structure and conventions of ancient medical practice reinforced, as well as reflected, social and cultural values.

The diagnosis of disease was made by taking a history and by thorough observation and examination of the patient's appearance, pulse, movement and excretions. In general, treatment was gentle and expectant, allowing the natural healing power of the body (*vis medicatrix naturae*) to exert its effect, aided by careful attention to diet, exercise, sleep and the environment, and only occasionally bolstered by more invasive therapies such as herbal remedies, massage, purging and blood-letting or cupping (see Figure 2). However, these secular treatments were often combined with religious rituals. When patients came to a healing temple, or *asklepion* (named after the Greek god of medicine, Asklepius), where many doctors were trained, they made sacrifices and presented votive offerings that were intended to appease the gods and restore health. In this way, the natural and holistic Hippocratic approach was blended with magico-religious, astrological and folk medicine.

Figure 2 Vessel for cupping (a form of blood-letting) discovered in Pompeii, dating from the first century CE (Wellcome Library, London)

The use of surgery and manipulation by Greek and Roman doctors is particularly interesting. We know from his writings that Galen performed surgery to aid the healing of partially amputated fingers and limbs or to remove foreign bodies from the head, eyes and stomach. Other surgical techniques included dentistry, the removal of cataracts and bladder stones, and the treatment of injuries sustained in battle. Archaeological evidence suggests that trepanning (or trephining), which involved drilling holes in the skull, was used, as in some indigenous South American cultures, to treat fractures. It may also have been employed in the treatment of epilepsy and mental illness. In addition, a particular manoeuvre used to reduce a dislocated shoulder has been attributed to Hippocrates, who is also credited with the invention of the Hippocratic bench, a device that applied traction to promote the healing of broken bones and correct curvature of the spine.

Both the Greeks and the Romans emphasized the importance of maintaining health and vigour through sport and hygiene. Following the successful establishment of the Olympic Games in the eighth century BCE, the Greeks built public parks, stadiums and gymnasiums, not only to provide facilities for athletes to train and compete but also to encourage the general population to keep fit and prevent obesity. In addition, Greek and Roman authorities introduced hot sweat baths and swimming pools to help improve fitness and relieve muscular aches and pains. In the Roman Empire, public baths allowed urban dwellers to cleanse their bodies, while aqueducts provided clean water for the emperor, private consumers (on payment of a tax) and military personnel. These initiatives were not only part of a widespread civic culture of cleanliness but were also linked to a growing obsession with grooming in order to improve appearance.

Hippocratic medicine was neither a monolithic nor unchallenged system of understanding and treating disease. Contemporary physicians contested and adapted many of the features of traditional Greek medicine. Galen, in particular, wrote prolifically on health and disease, sometimes confirming and sometimes disputing the teachings of Hippocrates. In fact, much of our knowledge and understanding of Hippocratic medicine is derived from Galen's commentaries on the Corpus and their subsequent translation into Latin, Syriac and Arabic. Influenced not only by Hippocratic humoralism but also by Aristotle's notion of the 'golden mean', sometimes interpreted to suggest 'moderation in all things', Galen advised patients to pay close attention to six 'necessary activities' to retain humoral balance and preserve health: moderating the air and environment; adjusting food and drink; ensuring sufficient and appropriate exercise and rest; promoting evacuation of urine and faeces; obtaining adequate sleep; and maintaining a healthy state of mind. Galen also dissected animals in order to develop a more refined account of human anatomy and physiology, expanded the range of herbal recipes used to prevent and treat disease, and extended the humoral system to include a fifth humour, pneuma, a vital principle that corresponded broadly with the ancient Chinese notion of Qi. As a result of his move to Rome, in approximately 162 CE, Galen transmitted both his own work and that of Hippocrates to the Roman Empire and beyond.

Many of the diseases that afflicted the inhabitants of ancient Greece and Rome were incurable. As a result, doctors focused particularly on preventative measures, such as moderation in food and exercise and improvements in personal and collective hygiene, and on effective pain relief, including the administration of opium, in the form of theriac, a concoction of fermented herbs in honey, or mithridatium, a mixture of over sixty ingredients initially developed by King Mithridates (134–63 BCE) as an antidote to poisons. In addition, doctors advised patients to adopt a stoical attitude to their condition. In his letters to Lucilius, a friend who was a civil servant in Sicily, the Roman philosopher and writer Seneca (4 BCE–65 CE) recommended comforting thoughts, the devotion of friends, determination and pleasurable distractions as the most effective means of enduring pain and disability. A doctor's reputation was dependent partly on their bedside manner and partly on the accuracy with which they could predict the outcome of disease, rather than on their capacity to cure. Patients with chronic, terminal disease sought refuge in dedicated sanctuaries that, like modern hospices, provided the opportunity for rest and reflection. The practice of medicine in ancient Greece and Rome thus incorporated a philosophy of life that not only helped patients to cope with disease but also reinforced personal, social and professional commitments to maintaining the stability, health and happiness of ancient populations.

East meets West

All ancient medical systems were to some extent distinct. Hippocratic humours corresponded only indirectly to the dynamic forces and constituents of the body that characterized Egpytian, Chinese and āyurvedic medicine. Nevertheless, it is possible to identify shared concepts and approaches among these traditions. Ancient Chinese, Indian, Egyptian and Graeco-Roman doctors and their patients understood disease primarily in terms of the balance and flow of elemental energies, as well as in terms of the impact of environmental change and the intervention of the gods. These ancient cultures also attempted to maintain and restore health in much the same ways, using herbs, lifestyle changes and occasionally surgery to promote fitness, relieve pain, remove tumours, limit the spread of infectious diseases and reduce the effects of chronic illness. In some cases, these would have been effective remedies.

How can we explain these similarities between parallel traditions? At one level, the resemblances may stem from comparable levels of knowledge about the body. In most ancient civilizations, dissection was prohibited or limited, leaving doctors to infer anatomical structures and physiological processes from surface examination and observation of the pulse and excretions. At another level, however, there were clearly processes of exchange between Eastern and Western medical cultures. Infectious diseases travelled via trade routes, thereby presenting ancient populations with similar clinical challenges. Chinese and Indian medicine percolated through other Asian countries, and Egyptian medicine had an impact on Graeco-Roman traditions through the migration of doctors (and their ideas and therapeutic knowledge) across regional boundaries. Thus, even the seemingly authoritative Hippocratic and Galenic approaches to medicine were derived from other traditions rather than being wholly innovative.

Ancient medical traditions were not static. In all cases, early theories and practices were modified over time according to social, political, religious and cultural beliefs. Nevertheless, it is clear that these traditions continued to shape experiences of ill health, as well as theories of disease, for several centuries. Humoral approaches to the body and its failings informed medieval understandings of health and disease in both the East and the West. Similarly, preventative and therapeutic interest in the role of lifestyle or regimen and the establishment of dedicated spaces for healing became increasingly important to medieval doctors and their patients. Rather strikingly, given the secular nature of most ancient medical systems, notions of health and the delivery of medical services also became increasingly entwined with, and configured by, faith and religion.

2
Regimen and religion: medieval medicine

How many valiant men, how many fair ladies, how many sprightly youths, whom, not others only but Galen, Hippocrates or Easculapius themselves, would have judged most hale, breakfasted in the morning with their kinsfolk, comrades and friends and that same night supped with their ancestors in the other world.

Giovanni Boccaccio, *Decameron*, 1353

The term medieval conjures up images of darkness, disease and death. Scarred by the harrowing consequences of pestilence and famine, by the religious tyranny and military terror associated with the Crusades, and by the social inequalities generated and sustained by feudalism, the Middle Ages have been portrayed as another country, far removed from our own secure world. While medieval populations were certainly plagued by anxieties about survival both in this world and the next, any assumption that the period between the collapse of the Roman Empire in the fifth century and the fall of Constantinople in 1453 was either one of intellectual stagnation or of decline from the zenith of the classical world would be misguided at a number of levels.

In the first instance, references to the 'Dark Ages' have focused habitually on western Europe and ignored the vibrant intellectual traditions that emerged in Byzantium and further east. In the early Middle Ages, between approximately 500 and 1100 CE, Islamic and Jewish scholars played a critical role in the translation and transmission of ancient texts and in the refinement of Graeco-Roman theories and practices of medicine. This process was not only important for the immediate continuation of Hippocratic and Galenic formulations of humoralism but also for the longer evolution of Chinese, Indian and Western medicine through the migration of people and the transmission of ideas across geographical and ideological borders. In the past, historians have also concentrated too much on the scholastic medicine that was fostered, and some would argue languished, in European monasteries and libraries. Preoccupations with learned medicine have diverted historical attention away from the variety of secular and religious healers and carers to whom the sick and suffering turned in times of distress. In doing so, historians have failed to acknowledge the manner in which, across the Middle Ages, faith and medicine did not necessarily compete but combined to generate new charitable and state institutions for the sick, poor and disabled.

Finally, we have perhaps too often emphasized the differences, rather than similarities, between health and medicine in the past and the present. Obsessions with illness and death, with the pain of intractable chronic disease, with the rise of mental illness and with seeking the advice of a wide range of state-sanctioned and alternative practitioners characterize our own age as much they do the Middle Ages. Indeed, the legacy of the Middle Ages is evident in various modern locations and customs: in the persistent reliance on regimen, that is, on adopting moderate balanced lifestyles, to prevent and treat disease; in the development of specialized hospital care dependent both on charitable donations and municipal support; and in the primacy of universities and teaching hospitals as places where students learn the art and science of medicine.

The medical landscape of the Middle Ages was fertile and fluid. Poverty and hunger dictated the patterns of acute infectious and chronic diseases, epidemics such as plague shaped the social and political fortunes of medieval communities, Islamic and Jewish scholars contributed to the perpetuation and adaptation of earlier medical theories and practices, and new institutions emerged for educating doctors and delivering health care. These historical transitions raise a number of questions about the medieval medical world. How did medieval populations experience and cope with disease? How was medical knowledge developed and transmitted? How and why did medicine change during the medieval period? What role did religion play in the delivery of health care to the sick and poor? And to what extent did developments in the Middle Ages provide the intellectual, practical and aesthetic foundations for Renaissance understandings and representations of the body in health and disease?

Poverty and pestilence

Epidemic diseases were not unknown in the ancient world. Indeed, pestilence may well have been responsible for the decline of Rome and the shift of political and commercial power to the Byzantine Empire. The plague of Athens during the war with Sparta in the fifth century BCE, the Antonine plague in the second and third centuries CE and the Cyprian plague, which struck in 250 CE and was named after the Christian bishop of Carthage, contributed to the gradual deterioration and demise of Greek and Roman civilization. Contemporary descriptions of these plagues, which appear to have been spread by armies and traders, describe the wasting, diarrhoea, exhaustion and rapid decline that proved particularly dangerous during times of famine and poverty and that afflicted and killed people of all social classes, including the Roman Emperor Marcus Aurelius (121–180 CE). However, it was arguably the Justinian plague, which was introduced into Constantinople by the movement of soldiers and the exodus of merchants from Egypt during the sixth century, that had the greatest social and political impact. Characterized by high fever and swollen glands leading swiftly to delirium and death, the Justinian plague slaughtered approximately 1,000 people every day, thereby weakening the strength and authority of the Roman Empire and facilitating the adoption of Christianity.

During the Middle Ages, increased travel further encouraged the spread of epidemic diseases throughout Eastern and Western Europe and beyond. Although archaeological and documentary sources do not reveal the precise nature or extent of these diseases, it would appear that medieval populations were often decimated by leprosy, tuberculosis, syphilis and plague, as well as being troubled by a range of chronic conditions such as diabetes, tumours, asthma, insanity, and disability caused by accidents. The appearance and manifestations of disease were explained in a variety of ways by medieval doctors and their patients: as the product of divine providence or displeasure; as the result of poor hygiene leading to pestilent fumes or miasmas; as the manifestation of humoral imbalances generated by improper lifestyles; as a result of astrological forces; or in some cases as a direct consequence of medical intervention. At the same time, however, it was evident to medieval communities that disease and death were more common when urban populations were over-crowded, poor and hungry.

Diagnosis of the wide range of acute and chronic medical conditions to which medieval communities were susceptible focused on listening to patients and their families, on clinical examination of the patient's appearance and pulse, on the inspection of excretory products such as urine and faeces, and on astrological or numerological calculations in which planetary alignments, numbers and dates were regarded as omens invested with singular significance. Both diagnosis and prognosis (the prediction of the outcome of disease) were thus shaped not only by direct observation of a patient's symptoms but also by an understanding of the natural and spiritual worlds and their regulatory forces. During the late Middle Ages, uroscopy (sometimes referred to as urinomancy or, more recently, urinalysis) developed as a prominent method for diagnosing and calculating the probable outcome of disease. First used by ancient Greek physicians, in the late Middle Ages uroscopy was increasingly advocated as a diagnostic tool by Byzantine, Hebrew and Western European physicians. Indeed, during the twelfth century, the French court physician Gilles de Corbeil (1165–1213) introduced a specially designed glass bottle, or matula, for examining urine.

The colour of the urine, the presence of sediments or surface bubbles and the smell and taste of the urine were thought to reveal specific forms of humoral imbalance and to signify distinct types of organic disease (see Figure 3). Diabetes, for example, was diagnosed largely on the frequency with which some patients urinated. According to the Greek physician Aretaeus, diabetes constituted 'a melting down of the flesh and limbs into urine', a feature that led later medieval and Renaissance authors such as Paracelsus (1493–1541) and Thomas Willis (1621–75) to refer to it as the 'pissing evil'. Excessive urination led physicians to taste the urine in order to determine its composition. As Willis noted, the urine of diabetics was known to be 'wonderfully sweet like sugar or honey'.

In ancient China, diagnosis had depended not only on physical examination but also on ritual and divination. Diviners interpreted signs and objects to identify the causes of illness, including ancestral curses, to predict the likely outcome of disease and to determine appropriate treatments, including sacrifices and offerings. In the medieval period, religious rituals remained important. Healing practices combined the use of acupuncture, moxibustion and herbal remedies with confession, meditation, moral restraint,

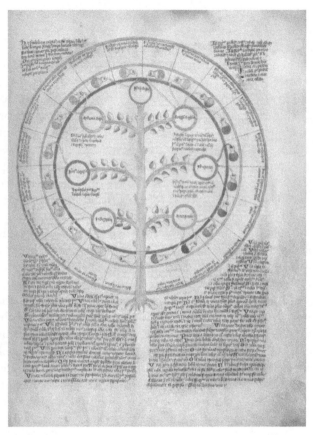

Figure 3   Text and illustration on 'urinomancy' or urine analysis (Wellcome Library, London)

worship and community work. Alchemy was used to gain physical and spiritual purity. Both male and female priests and healers could offer spiritual advice in relation to health, and prominent physicians became influential figures in governments and courts. Although Daoism was challenged by Buddhism in the provision of medical services, a Daoist sect known as the Celestial Masters largely governed the practice of medicine and dictated political authority in China during the Middle Ages.

The condition that perhaps best captures the perils of life, as well as the fear and stigma of disease in the early Middle Ages, is leprosy. Although the perceived causes of leprosy varied from place to place and across time, descriptions of the physical deformities of lepers, as well as efforts to segregate and marginalize them, were relatively consistent. Disfigured by scaly flesh, by ulcerated and paralysed hands and feet, and by facial defects and foul-smelling breath, lepers were regarded as undesirable outcasts who were capable of infecting healthy populations. According to one thirteenth-century leper's lament, for example:

'Nobody will want to look at me,
Nobody will dare to speak to me,
Because I have polluted breath.'

The physical pain of leprosy was accentuated by social disapproval and moral condemnation, particularly because leprosy was often thought to be God's punishment for promiscuity and because the isolation of lepers was supposedly sanctioned in the Bible. Lepers were forced to carry bells, clappers or horns to warn neighbours of their presence and were sometimes ordered to leave the village, town or city in which they had lived and worked for many years. The husband or wife of a leper could obtain a divorce once the disease had been confirmed by a jury of doctors, priests and civic administrators, in a procedure sometimes referred to as the 'judgement of lepers'. Because of the personal and social consequences of being identified as a leper, some patients and their families disputed a diagnosis of leprosy, which was indeed often difficult to prove. Early European hospitals for lepers, known as leprosaria or lazarettos, were intended not only to provide food and shelter for the infirm but also to ensure their safe removal and long-term segregation from respectable society. The absence

A LEPROUS MAN

Then the priest shall examine him, and if the diseased swelling is reddish-white on his bald head or on his bald forehead, like the appearance of leprosy in the skin of the body, he is a leprous man, he is unclean. The priest must pronounce him unclean; his disease is on his head.

The leper who has the disease shall wear torn clothes and let the hair of his head hang loose, and he shall cover his upper lip and cry, 'Unclean, unclean!' He shall remain unclean as long as he has the disease; he is unclean; he shall dwell alone in a habitation outside the camp.

Leviticus, 13: 43–46

of an effective cure, other than the hopeful incantation of prayers or in some cases the amputation of diseased limbs, consigned people diagnosed with leprosy, and indeed their families, to lives of isolation and despair.

By the fourteenth century, leprosy was on the decline. New diseases, however, rapidly emerged to take its place as sources of sickness and dread. The most infamous late medieval scourge was the plague, which appeared in Europe in the 1340s. At the time, it was known as the 'Pestilence' or the 'Great Mortality' and was only later referred to as the 'Black Death'. Probably imported by traders travelling from China along the Silk Road, within only a few years plague had killed approximately twenty million people, or between thirty and sixty percent of the European population. Doctors and patients explained the origins and spread of plague in various ways; the disease was primarily regarded as the product of sin or as the result of unhealthy and immoral lifestyles leading to humoral imbalance. In addition, the plague was blamed on the miasmas or 'pestilential atmospheres' generated by poor sanitation and unburied bodies. Although the epidemic killed people from all social classes, it was known to be more treacherous during times of poverty and famine. The higher incidence of plague deaths amongst the lower social classes may have been linked not only to poor nutrition and hunger but also to the ease with which the rich were able to flee from affected cities into healthier rural environments.

Contemporary descriptions of the plague, particularly those by the French surgeon and papal physician Guy de Chauliac (1300–68) and the Italian writer Giovanni Boccaccio (1313–75), highlight the ferocity and rapidity with which patients were seized by fever and the eruption of abscesses or buboes, initially in the groin, under the armpits or on the neck but subsequently spreading throughout the body. Death was swift and painful, leaving families and communities devastated. The highly contagious nature of the condition led to fear and segregation: 'The father did not visit his son, nor the son his father', wrote de Chauliac in 1363. 'Charity was dead and hope destroyed.'

The Black Death carried profound socio-economic and political consequences. Across Europe, the decline in population led to a reduction in trade and to rising wages for a greatly depleted workforce. Geographical and social mobility were enhanced by the opportunities for workers to travel to regions where wages were higher. Social disruption led to greater violence, evident in the English Peasants' Revolt of 1381, when rebels ransacked churches and manors and murdered religious and civil leaders. Such events initiated the gradual decline of feudalism, as labourers began to resist the government's ruthless enforcement of feudal service. At the same time, the plague served to curb the power of the Church, partly because many priests died during the epidemic but also because, more concerned with souls than bodies, religion was seen to be ineffective in relieving symptoms or preventing the spread of sickness and death.

Although plague proved to be incurable, several preventative measures were introduced. Crosses were placed on the doors of infected houses and on the bodies of plague victims, to identify potential sources of contagion (see Figure 4). The inhabitants of

Figure 4   Mortuary crosses placed on the bodies of plague victims, c. 1348 (Wellcome Library, London)

cities were forbidden from visiting plague areas and no corpses were allowed to be buried within the city walls. Temporary isolation, or quarantine, was introduced in order to detect asymptomatic carriers of the disease and people wore flowers round their nose to protect them from harmful emanations. Religious ceremonies and superstitious rituals were also employed: communal prayers and the performance of penitential rites, participation in confession and the Eucharist, visits to healing shrines and the possession of talismanic objects, such as engraved rings and stones, offered consolation and hope, if little genuine prospect of protection. Subsequent outbreaks of plague continued and were recorded in plague tracts and in the mortality statistics that many parishes introduced to chart the natural history of the condition.

As plague receded after 1450, other infectious diseases became more prominent. Medieval populations, particularly children, women and city-dwellers, were increasingly troubled by tuberculosis, influenza, syphilis and dysentery, many of which were described in detail for the first time during the late medieval period. During the late fourteenth and early fifteenth centuries, for example, the characteristic genital lesions of syphilis (or 'the Great Pox') were first identified and referred to as a venereal disease; that is, as a condition transmitted through sexual contact, although many people continued to regard it as a punishment for sin. One theory accounting for the appearance of syphilis in this period is that the disease was imported into Europe in the 1490s following the return of Christopher Columbus (1451–1506) and his sailors from the Americas. However, the disease may already have been present in Europe, particularly in Italy, before this time and been spread by French soldiers who had been fighting in Naples: for this reason, it was often referred to by Italian authors as the 'French disease' ('*morbus gallicus*') and by French writers as the 'Neapolitan disease'. Some treatments for syphilis were introduced in this period, including the application of ointments and powders containing guaiacum or mercury, which remained in use until the twentieth century.

Already old by the age of 45, adults were also afflicted by chronic conditions such as cancer, gout, asthma, diabetes, physical injuries and disabilities, skin diseases, epilepsy, rotten teeth and gums, and problems associated with menstruation, pregnancy and childbirth. Like their infectious counterparts, these diseases were exacerbated by poor living conditions, poverty and starvation and were treated with combinations of herbal remedies, dietary ingredients, charms and prayers. Treatments were carefully adjusted according to the age and sex of the patient. Western European doctors also adopted and adapted the innovative surgical techniques of their Arabian counterparts. The English surgeon John Arderne (c. 1307–80), for example, was renowned for operating on patients with an anal fistula, borrowing the surgical procedure from Arabic sources and describing the techniques in detail in books originally written in Latin but rapidly translated into English.

Although physical illness and pain dominated people's lives, various forms of persistent and recurring mental illness were also prevalent. For some religious commentators, the rapturous visions of saints and martyrs were the product of ecstatic revelation. But for many patients and their doctors, the causes and consequences of insanity were more mundane and dispiriting. Thought to be linked to the phases of the moon, 'lunacy' was usually regarded as the result of satanic possession or divine displeasure precipitated by sexual immorality and over-indulgence. Medieval treatments for madness included fasting and prayer, as well as purging, bleeding and trephining to release dangerous demons and humours. For many patients these treatments were ineffective and madness could lead to prolonged anguish, despair and sometimes suicide. In cases of suicide, a diagnosis of insanity was often beneficial for the family. If a person who committed suicide or self-murder was judged to be sane and therefore guilty of a crime, his or her possessions could be confiscated by the Crown, but if the person was deemed to 'lack sense or reason' their family would be permitted to inherit the estate.

Medicine and faith

Medieval approaches to health care were based loosely on earlier secular traditions that were gradually modified by religious customs. In the West, Graeco-Roman humoralism was merged with Christian ethics. Although to some extent Christian leaders subordinated secular learning for their own purposes, medicine and faith were often closely linked in theory and practice, as both priests and doctors preached or prescribed moderation in consumption and behaviour. The central role of Christianity in healing was legitimized by the belief that Jesus Christ was the 'one physician' capable of healing both body and soul, a notion embodied in the term 'Christus medicus', which was propagated in particular by the writings of St Augustine (354–430 CE). In addition, refinements of classical medicine were forged by Islamic and Jewish scholars adopting and adapting Hippocratic and Galenic principles.

Eastern approaches to health and disease were similarly infused with religious values. Chinese approaches to illness and healing were strongly influenced by Daoism and Buddhism, particularly during the Sui (581–618) and Tang (618–907) periods. With an emphasis on reincarnation, Buddhism advocated prayers and meditation as the primary means of achieving spiritual, mental and physical health and Buddhist monasteries provided shelter and sustenance to Asian populations coping with the ravages of epidemic diseases. A combination of magico-religious and secular healing practices similarly characterized medicine in Japan and Korea between the sixth and sixteenth centuries, as well as the system of Tibetan medicine as it emerged from the assimilation of Indian and Chinese traditions during the seventh century. As the case of Tibet demonstrates, these healing traditions were not entirely distinct, since the movement of religious figures, healers, patients and medicines served to provide opportunities for unfamiliar practices to shape local formulations of disease and the delivery of medical care.

After the collapse of the Roman Empire and the rise of Byzantium and Arabia as intellectual, political and military forces, the texts and teachings of classical Greek doctors were preserved in medieval monasteries and libraries. The transmission of humoral medicine was dependent initially on the educational commentaries of writers such as Oribasius (c. 320–400), Stephanus of Athens (seventh century) and subsequently on the work of Byzantine, Islamic and Jewish scholars, who refined the ancient approaches embedded in the treatises of Hippocrates and Galen and in the encyclopedia of herbal remedies compiled by Dioscorides. The principal architects of medieval medicine in Europe and the Middle East were Abū Bakr Muhammad B. Zakāriyya' al-Rāzī (865–ca. 925); the Arabic physician Abu al-Qasim al-Zahrāwi (936–1013); Ibn Sīnā (980–1037); 'Abū l-Walīd Muhammad Ibn 'Ahmad bin Rušd (1126–98); and the Jewish scholar Rabbi Moses ben Maimon (1138–1204). In the past, historians often westernized the names of these scholars and physicians, referring to Rhazes, Albucasis, Avicenna, Averroes and Maimonides respectively, but it is now regarded as more accurate and respectful to use their original Arabic or Hebrew names.

Born in Iran, al-Rāzī completed his medical education in Baghdad before practising medicine and directing hospitals, first in his home town of Rayy (near present-day Tehran) and subsequently in Baghdad. His extensive writings not only shaped the development of medicine but also contributed widely to the study of philosophy, alchemy, mathematics, theology and music. Like many earlier Greek physicians, al-Rāzī rejected religious approaches to knowledge and truth and preferred to base his understanding and treatment of disease on reason and experience. Although some of his work constituted compilations of Greek, Syrian, Indian and Arabic medical knowledge, he was an independent and critical thinker who challenged established intellectual authorities. In publications such as *Kitāb al-Hāwī fi't-Tibb*, known as *The Comprehensive Book of Medicine,* al-Rāzī questioned some of the interpretations of Galen and aligned himself more directly with Hippocratic approaches to the observation, diagnosis and treatment of disease. In addition to writing on the principles and practice of medicine, including many works dedicated to local rulers and princes, al-Rāzī's extensive clinical experience was also compiled into a casebook, which was subsequently used as a student textbook.

Approximately one century later, Ibn Sīnā, arguably the most influential medieval medical author, published his *Kitāb al-Qānūn fī-t-Tibb* (*Canon of Medicine*). Born in Afshana, near Bukhara in Uzbekistan, Ibn Sīnā produced nearly 300 works on mathematics, law, theology, philosophy and medicine. Comprising five books, which focused in turn on the general principles of medicine, simple remedies, anatomically specific diseases, systemic conditions and the medicinal use of compound drugs, Ibn Sīnā's *Canon* was translated into Latin by Gerard of Cremona (1114–87) in the late twelfth century and became a standard educational text for medical students in Italy during the Renaissance and early modern period. The *Canon* reflected ancient and medieval preoccupations with regimen and incorporated a combination of approaches, which included purging, vomiting and blood-letting, as well as the use of foods and herbal remedies administered either by mouth or inhalation, to restore humoral balance and relieve the symptoms of both acute and chronic disease.

The works of Rabbi Moses ben Maimon were also crucial to the survival and refinement of ancient medicine. Born in Córdoba in Spain in 1138, he gained fame as a doctor and philosopher, becoming court physician under Saladin the Great in Cairo and the leader of the Jewish community in Egypt. His publications, which were written in Arabic but often translated into Hebrew, included commentaries on the works of Galen and Hippocrates; medical texts on specific activities and conditions, such as asthma, sexual intercourse, haemorrhoids, poisons and fits; a collection of medical aphorisms; a pharmacopoeia of medicinal drugs; and a wide range of influential theological and philosophical reflections. Particularly prominent amongst his publications was his synthesis of medieval medicine, *On the Regimen of Health*, which provided a broad introduction to the maintenance of health and the prevention of disease, and a more focused treatise on asthma, which was translated at least three times into Hebrew during the fourteenth century and twice into Latin in the fourteenth and fifteenth centuries.

Like their ancient predecessors, and to some extent like their contemporary Eastern counterparts, medieval authors in the West proposed a system of medicine that prioritized a healthy lifestyle and an appropriate diet. Echoing Galen's emphasis on the necessary activities needed to prevent or treat disease, medieval doctors highlighted the importance of paying careful attention to what were generally known as the six 'non-naturals' (air, food and drink, movement and rest, emotions, sleep, and excretion) and to the frequency of sexual intercourse to preserve and restore health or in some cases to promote fertility. In addition to monitoring and rebalancing mental and physical health through analysis and modification of regimen, medieval medicine included the use of herbs to re-establish the appropriate humoral equilibrium. Rabbi ben Maimon's advice to patients with chronic diseases such as asthma, for example, included recommending the consumption of a wide range of foods, including fresh fruits and barley broth, intended to reduce the accumulation of harmful humours. He also emphasized the importance of avoiding emotional disturbance and adopting a positive attitude in order to reduce pain and manage illness: sadness, fear, worry, anger and distress, he argued, could reduce appetite, deplete energy and hasten death.

As the works of al-Zahrāwi demonstrate, many medieval doctors also possessed considerable surgical skills, including the use of cautery to heal abscesses and wounds and remove tumours; the application of sutures; and the operative treatment of fractures, bladder stones and obstetric conditions. Archaeological evidence suggests that much surgical experience was gained during the Crusades. Sometimes carried out in field hospitals, emergency surgery was necessary to treat the blade injuries, penetrating arrow wounds, head injuries, mutilations and burns that were inflicted either during battle or as a form of torture. Scalpels, forceps, scissors, saws and cautery irons have all been recovered from medieval settlements. In addition, soldiers on both sides during the Crusades were sometimes treated surgically for bone infections, haemorrhoids and the accumulation of fluid in the abdomen, and blood-letting was regularly employed by military surgeons. Pain during surgery or as the result of disease was managed with opium or less effective analgesics such as mandrake, hemlock and deadly nightshade.

Reaffirming the combination of sacred and secular approaches that were adopted to prevent the spread of epidemics, medieval authors endorsed a similar array of spiritual and corporeal remedies for chronic diseases. In some cases, penance was prescribed to atone for sins, alongside herbal remedies. This is not to say that medicine and religion always operated in harmony: for many, cure of the soul remained more important than cure of the body. Nevertheless, both religious and secular formulations of disease continued to influence the theory and practice of medicine in western Europe and also had an impact on Eastern traditions. In India, for example, the incorporation of Galenic medicine from Islam during the eleventh century led to the emergence of a novel medical tradition, ūnānī tibb, which to some extent challenged more established āyurvedic forms of practice. In addition, during the Middle Ages various aspects of medicine were institutionalized. Using the compilations of ancient texts produced by medieval scholars, students increasingly learned their craft in universities as well as in apprenticeships with established practitioners. Similarly, although consultations continued to take place in the home, patients were often examined and treated in the hospitals that were being built in the rapidly expanding cities and commercial centres.

Learning and healing

While early medieval learned medicine was developed primarily in monasteries and libraries, during the late medieval period (that is, between approximately 1100 and 1500) medical education and practice also became associated with universities and hospitals, institutions that were made possible largely through the preservation of key texts and the transmission of ancient knowledge by religious scholars such as al-Rāzī, Ibn Sīnā and ben Maimon. It was through these institutions that novel hierarchies of practice were forged, according to which different types of medical practitioners were separated by training, status and money. New approaches to medicine also emerged, particularly in relation to the growing reliance on anatomy.

A wide range of medieval doctors and healers attended the sick, supported families and communities, and provided health care. At the summit of an emerging hierarchy of practitioners were the physicians, who were increasingly separated from their peers by a university education that trained them in the learned traditions of Graeco-Roman medicine and equipped them for elite practice in cities, monasteries and courts. The education of physicians, who were regarded by the Italian philosopher and theologian Thomas Aquinas (1225–74) as the 'makers of health', was made possible by the creation and expansion of universities, such as those at Bologna (founded c. 1088), Salerno (1100), Paris (1110), Oxford (1167), Montpellier (c. 1181) and Cambridge (1209), where students were taught natural science and philosophy as well as medicine. By the late fifteenth century, there were approximately fifty universities across Europe. These institutions provided Latin translations of key medical treatises as well as offering instruction in innovative approaches to medical knowledge of human form and function based on the dissection of cadavers. According to some historians, this growing focus on the demonstration of anatomy constituted a late medieval renaissance of classical learning and the origins of modern medical education.

Below the physicians in terms of status and economic reward were the surgeons, who were trained not only as healers but also in this period as barbers, a pairing of trades that was reinforced in Europe by the creation of professional guilds, such as the English Guild of Surgeons, founded in 1368, and the Guild of the Barbers of London, established in 1462, which merged to form the Company of Barber-Surgeons in 1540. Surgical

training was through apprenticeship, during which students
developed manual skills and learned about the use of instruments
rather than immersing themselves in classical texts. Proficiency in
surgery equipped young doctors for careers in the army or navy.
Although there were often professional rivalries between physicians and surgeons, it was well recognized that to provide the best
care practitioners needed to be conversant with the principles of
both general medicine and surgery.

At a more local level, a variety of healers provided both naturalistic and religious systems of medical care in a complementary,
rather than antagonistic, manner. People with incurable conditions in particular turned to religion for comfort: the use of healing shrines, penance, pilgrimage and prayers to specific saints for
miracles reflected, as well as reinforced, spiritual understandings
of health, disease and death and were intended to encourage
endurance and resilience in the face of the transience of earthly
life. With death ever-present, approaches to death and dying
constituted an important part of life. During the late Middle
Ages, advice on how to recognize the signs of imminent death
or on how to prepare individuals and communities for death was
incorporated into visual depictions of the 'dance of death' (*danse
macabre*) and into textual descriptions (*ars moriendi*) of how to die
well (see Figure 5).

The practice of medicine was not restricted to men. Women
also offered medical advice and services, such as nursing and
midwifery, not only in the community but also in hospitals,
monasteries and lazarettos. In many places, the practice of
midwifery in particular was regulated by both the Church and
the state and supervised by male doctors. The status of women
healers was ambiguous but some, such as Hildegard of Bingen
(1098–1179), became famous partly through their clinical practice and partly through their writing, which sometimes drew on
their own experiences of health and illness. Born in Germany,
Hildegard entered a monastery, where she began to write not
only theological commentaries and reflections on spiritual

Figure 5   The 'dance of death' or *danse macabre*, 1493 (Wellcome Library, London)

experience but also musical scores, biographies of saints and medical texts. Her clinical works focused on the functions of different parts of the body, on methods of diagnosis, and on the use of plants, animals and minerals, as well as cupping and bleeding, to treat specific conditions such as headaches, coughs, pain, diarrhoea, infertility and impotence. According to Hildegard, stomach pains could be cured with a mixture of peony, thistle, ginger and pepper, while a woman's fertility could be restored with the cooked womb of a cow or sheep. Hildegard's medical career exemplifies the fluidity and complexity of late medieval medicine: her practical training in the monastery's infirmary, her knowledge of classical texts obtained from the monastery's library, her experience of treating sick patients and her spiritual convictions together provided the conceptual and practical basis for her recipes for maintaining health and treating disease.

The creation of new sites of learning was mirrored by the establishment of novel places of healing. From the twelfth century onwards, hospitals became increasingly important in the delivery of support and treatment for the infirm. Of course, institutions of this nature were not new: in the ancient world, healing temples had offered both education and medical care. But from the late medieval period, both charitable and municipal hospitals, often modelled on earlier Byzantine institutions, were built to treat soldiers and the poor or to offer relief and sustenance to pilgrims and other travellers (see Figure 6). In addition to providing medicines and some surgery, hospitals offered food, rest and nursing

Figure 6   Caring for the sick in the Hospital of Santa Maria della Scala, Siena (Wellcome Library, London)

care to patients, and new spaces for the education of medical students. St Bartholomew's Hospital in London was founded in 1123 and St Thomas's Hospital, named after the Archbishop of Canterbury Thomas à Becket (1118–70), was established in 1215. In Islamic regions, prominent physicians such as al-Rāzī presided over hospitals as places of both healing and learning.

Within the context of growing concerns about the spread of epidemic disease, specialist hospitals were also designed to provide a sanctuary for lepers or the victims of plague and to protect the rest of society from the threat of contagion. In some cases, these institutional strategies for controlling the spread of infectious diseases developed into wider public health measures, including the compilation of mortality statistics (in 'books of the dead'), the imposition of quarantine and restrictions on travel, and the introduction of more elaborate methods for monitoring the burial of infected corpses. As the spectre of leprosy receded, fears about the disruptive social effects of uncontrollable mental illness deepened. Empty lazarettos and new hospitals became places of asylum for the insane. Although many people with mental illness remained in their communities, cared for at home or by neighbours, from the fifteenth century onwards those deemed insane were increasingly committed to hospitals, such as those at Valencia, Seville, and Bethlem in London. Isolated and segregated, the mentally ill replaced lepers as the most visible and perhaps most feared social outcasts.

It would be a mistake to regard specialized medieval institutions and the associated systems of medical surveillance and treatment as self-evident precursors of modern hospitals. Treatment in a medieval hospital was not merely a matter for doctors but was linked to the hospital's capacity to provide solace, comfort, rest, food, nursing care and spiritual guidance for the poor, sick and infirm. In addition, there was no smooth pathway from the medieval to the modern hospital. During the Black Death, high death rates within institutions led to declining support for hospitals. In many places, evidence of poor management and low standards of care resulted in calls for hospital reform. Rising concerns about the power of the clergy and their inability to combat epidemic disease coincided with the dissolution of the monasteries and forced many European hospitals to close or adapt themselves to new social and political pressures. In spite of these trends, it is noticeable that the number of hospitals increased over time: in England, for example, the number of hospitals rose from fewer than 70 in 1080 to approximately 600 by 1530.

In the West at least, the changing form and purpose of medieval hospitals or late medieval universities were part of a general transformation of European society and culture, a process that was closely linked to the rise of urban centres, the spread of disease, the creation of professional guilds and corporations, and the shifting authority of religion. As the shape of medical institutions began to change in response to these developments, certain practitioners, particularly religious healers and women, were increasingly marginalized from hospital work. Although focused primarily on training doctors and dispensing health care and support to sick and impoverished populations, medieval institutions for healing and learning were also the product of complex interactions, and sometimes competition, between the interests of the state, the Church and the medical profession.

Translation and transmission

During the Middle Ages, medical theory and practice constituted a more diverse and dynamic form of medicine and health care than the term 'medieval' initially suggests. Ravaged by both acute epidemic and chronic diseases, contemporary populations in both the East and the West turned to a variety of secular and spiritual healers in the hope of immediate recovery and eternal salvation or reincarnation. Clinical knowledge owed much to the translation and dissemination of classical Graeco-Roman texts by Byzantine, Islamic and Jewish scholars. Medieval systems of diagnosis and treatment continued to be influenced by humoral theory or understandings of bodily balance but were often adapted according to local custom and need and interpreted in the light of emerging religious traditions, including Christianity, Islam and Buddhism. Initiated by charitable and philanthropic initiatives, the rise of late medieval institutions for education and healing provided the basis for a clearer hierarchical separation of physicians from surgeons and other practitioners, as well as a growing reliance on anatomy as the essential basis of knowledge for the educated doctor.

The fall of Constantinople in 1453 triggered a more pronounced revival of Greek culture and the rediscovery of Greek and Roman texts throughout western Europe and the Middle and Far East. Transmitted through Hebrew and Arabic translations, many ancient medical and philosophical works became available in Latin printed editions for the first time and were sometimes bound together into a single textbook, known as the *Articella*, designed for the increasing numbers of students entering European, and particularly Italian, medical schools during the fifteenth, sixteenth and seventeenth centuries. In addition to Latin translations of medieval treatises, the compilations and commentaries of major classical authors, such as Galen, Aretaeus and Celsus, were available in Latin from the late fourteenth and early fifteenth centuries. However, while ancient Greek, Egyptian, Arabic and Hebrew formulations of health and disease certainly continued to influence approaches to medicine, new forms of scientific research emerged, along with greater awareness of alternative healing traditions and novel resources for learning the principles of medical practice.

During the Renaissance, further developments in medical education and the delivery of health-care services were encouraged by the rise of anatomical dissection, associated with fresh approaches to the artistic representation of the human form; the spread of the printed book, the exchange of ideas and remedies between the East, Europe and the New World; and the emergence of new forms of scientific knowledge. Although many poor people remained untouched by learned medicine, relying instead on local lay knowledge to maintain and restore health and well-being, these intellectual, social and cultural processes served gradually to challenge ancient authority and to establish a conceptual and practical basis for novel understandings of disease and innovative approaches to medicine.

3

Bodies and books: a medical Renaissance?

> I began to think myself whether it had a sort of motion as if in a circle, which afterwards I found to be true, and that the blood was thrust forth and driven out from the heart through the arteries into the flesh of the body and all the parts, by the beating of the left ventricle of the heart...and that it returns through the veins into the vena cava, and to the right ear of the heart
>
> William Harvey, *De motu cordis*, 1628

The greatest revolution in human history took place in the middle of the fifteenth century. Various rudimentary mechanisms for printing books had been introduced in the late medieval period, most notably in Korea and China, where Bi Shēng (990–1051) and others had developed a system that was initially based on fragile moveable clay types but was subsequently improved by using metal and wood. The key innovation, however, was made by a German blacksmith and goldsmith, Johannes Gutenberg (1398–1468), who invented a moveable type printing press in around 1439. Having perfected the system, first in Strasbourg and then in Mainz, Gutenberg produced his most famous publication, the 42-line Gutenberg Bible, in 1455. Although illustrations and red-letter headings continued to be added to the texts by hand, the printing press greatly accelerated the production of books, initiating what historians have regarded as a printing revolution. As Gutenberg's technology spread, it carried with it the capacity to produce books more rapidly and cheaply, to disseminate literacy and learning more widely, to reacquaint people with ancient wisdom, to generate a transformation in scientific method, and to contribute to a gradual decline in the power and influence of the Catholic Church.

Of course, the introduction of printing was not the only factor in generating new patterns and forms of knowledge and authority. Even before Gutenberg's invention, European artists challenged medieval modes of representing the human form by reinstating classical aesthetics. In 1453, the fall of Constantinople accelerated this process of creative and intellectual revision by shifting the centre of learning away from Byzantium, first to Italy and subsequently to the principal cities of northern Europe. Nowhere was this trend more evident than in the realms of science and medicine. During the fifteenth and sixteenth centuries, the universities and medical schools at Padua, Bologna, Montpellier, Paris and elsewhere provided the academic and practical foundations for novel understandings of anatomy, as well as innovative approaches to health and disease. Increasingly, the opportunities offered by the Italian style of medical education attracted overseas students and doctors, who were eager to acquaint themselves not only with the insights of ancient medical authors, available for the first time in Latin translations, but also with contemporary techniques in dissection and comparative anatomy. On their return home, these visiting doctors helped to transmit Italian culture and scientific knowledge eastward and westward.

These developments are generally regarded as part of the Renaissance, a term first used by the painter and writer Giorgio Vasari (1511–74) in his biographical portraits of contemporary artists. The Renaissance refers either to a period of time or a cultural movement that occurred in western Europe between the fifteenth and seventeenth centuries, during which philosophers,

artists, musicians, scientists and doctors, amongst others, attempted
to engineer a substantial shift away from medieval mentalities and
endorsed ancient humanist approaches to art, science and medicine. Inspired by nostalgia for the beauty and symmetry of the
classical world, Renaissance humanists were committed to restoring the sanctity and purity of Greek culture in order to provide
firmer intellectual foundations for explaining and manipulating
the modern world. As several historians have suggested, Renaissance humanism was therefore a form of Hellenism that drew its
inspiration from classical approaches to aesthetics and emotion.
Within natural philosophy and medicine, this allegiance to
Graeco-Roman authority became manifest in the robust, if occasionally contested, return to the authentic teachings of Aristotle,
Hippocrates and Galen.

There is some truth to the assertion that the Renaissance
constituted, or at least initiated, a radical rupture in the fabric
of human life. Through a series of spectacular developments in
art and science, Renaissance scholars and artists generated fresh
scientific insights into the structure and function of the human
body. Yet the case for cultural change over continuity can be over-
stated. On the one hand, Eastern approaches to art and medicine
remained largely unaffected by intellectual and artistic developments in Europe. On the other hand, advances in scholastic or
learned medicine and changes in visual accounts of the body
only minimally affected health care either at home or in hospitals. Domestic and community approaches to health and sickness continued to be shaped by oral and scribal traditions loosely
based on humoral formulations of the body but largely distinct
from broader cultural shifts in medical theory.

Between the fourteenth and seventeenth centuries doctors
and scientists at first reinforced, but subsequently began to reject,
traditional forms of medical education, knowledge and practice
– processes that gradually changed the lives of patients and their
families. Dissection and anatomy played a pivotal, if sometimes

disreputable, role in promoting greater confidence in surgery. New approaches to evidence and proof generated during the Scientific Revolution transformed accounts of human physiology, particularly evident in the work of William Harvey (1578–1657) on the circulation of the blood, in the development of mechanical, rather than humoral, accounts of the sick body, and in the invention of instruments such as the microscope. By exploring the range of illnesses from which early modern populations suffered and the variety of approaches that were adopted to maintain or restore health in both the East and the West, we can begin to evaluate historical claims that the Renaissance and early modern period constituted a watershed in the history of modern medicine, one that provided relatively firm foundations for the rational scientific medicine that emerged during the eighteenth and nineteenth centuries.

Art and anatomy

The Renaissance is most often associated with developments in art and architecture. During the fourteenth century, Italian artists such as Duccio di Buoninsegna (1255–1319) and Giotto di Bondone (1267–1337) were challenging the ornate gothic style that had become popular during the later Middle Ages and cultivating new ways of depicting the human figure, with greater realism and an enhanced sense of emotion. Many early Renaissance paintings were frescoes of sacred figures created to adorn the ceilings and walls of chapels and churches, such as those in Florence and Siena. Although some artists, such as Sandro Botticelli (1445–1510), whose *Birth of Venus* became one of the most famous Renaissance images, initially resisted the move to realism, many fifteenth-century painters adhered to the classical style re-established by Giotto and his peers. As Renaissance artists strove to recreate the harmony, beauty and sculptural forms of classical art and architecture, they also rediscovered older techniques, including the use of egg yolks to bind coloured powders, a practice known as the 'egg tempera' method.

The most well-known Italian masters were Leonardo da Vinci (1452–1519), Michelangelo Buonarroti (1475–1564) and Raffaello Sanzio da Urbino (Raphael) (1483–1520). Together, these artists continued the early Renaissance preference for painting sacred figures, often nude, in idealized rural settings. Partly facilitated by the capacity of printing to reproduce woodcut illustrations as well as text, this approach to art spread to the cities of Northern Europe, where Jan van Eyck (1390–1441) and Albrecht Dürer (1471–1528) were similarly obsessed with religious themes and the new naturalistic commitment to truth and detail. Not all Northern European artists subscribed to classical modes of representation or religious iconography. Hieronymus Bosch (1450–1516), for example, developed an idiosyncratic style that anticipated the much later aspirations of surrealism. In addition, the Reformation particularly served to marginalize Catholic imagery. In spite of regional, religious and temporal differences, however, it is clear that most Renaissance painters and sculptors displayed a common devotion to restoring classical perceptions and depictions of the human body.

Drawing deliberately on the aesthetic norms of Greek art, Renaissance artists used nature as their guide. Careful study of nature (and particularly the body) was expected to provide the knowledge required for accurate portrayal of human form and emotion. In this way, familiarity with anatomy became a prerequisite for achieving the Renaissance ambition of realistically representing human figures. Initially, artists acquainted themselves only with surface anatomy but, increasingly, Leonardo da Vinci and Michelangelo, amongst others, began to dissect human corpses to uncover the arrangement of internal structures. As da Vinci's detailed drawings make clear, Renaissance anatomical knowledge of the muscles, bones and internal organs was extensive and incorporated not only into creative compositions but also into illustrations for medical texts sometimes compiled for rich patrons. Careful attention to the contours and fabric of the human body was not confined to the West. Doctors and artists from the Islamic Middle East, as well as from China and India, employed dissection of animals and humans for the express purpose of displaying more accurately and understanding more fully the anatomical arrangement and physiological purpose of different parts of the body.

Given the moral prohibition on human dissection in many ancient cultures, as well as widespread fears about the spread of infection from polluted corpses, Graeco-Roman and Egyptian insights into anatomy and physiology had largely been derived from the preparation of bodies for burial, from surgery or from studies of animals. Galen's accounts of circulation and respiration and his descriptions of the purpose of specific organs, for example, were based on dissections of frogs, sheep and pigs. It was because he had drawn his conclusions from anatomizing frogs that Galen, like Aristotle, insisted that the heart contained only three chambers. Although the bodies of murder victims were occasionally examined to discover the cause of death, ancient understandings of anatomy persisted during the medieval period, as students and doctors relied predominantly on the works of Hippocrates and Galen to direct their theory and practice. In the later Middle Ages and early Renaissance, however, this began to change, as several European universities introduced anatomy demonstrations for medical students, first in Bologna in the early fourteenth century and subsequently in other leading Italian and French medical schools. Artistic and medical interest in anatomical dissection served to subvert Galenic accounts of the body and to challenge the schematic illustrations of human form that had dominated medieval manuscripts.

Most scholars agree that one of the defining moments in the history of anatomy was the publication of an illustrated anatomical textbook by Andreas Vesalius (1514–64). Born in Brussels and trained in humanist medicine, Vesalius was a disciple of Galen. Having learned the principles of dissection in Paris, he travelled first to Louvain and subsequently to Padua where, in spite of the fact that he was a physician, he was appointed as a lecturer in surgery and anatomy. From the late 1530s, Vesalius produced a number of illustrated and educational texts for medical students, including depictions of anatomical structure, commentaries on the works of Galen and al-Rāzī, and discussions of the theory and method of venesection (blood-letting). However, it was the publication of *De humani corporis fabrica* (translated as *On the Fabric of the Human Body* and usually referred to simply as the *Fabrica*) in 1543 that established Vesalius as the leading exponent of modern medical anatomy.

The *Fabrica* consisted of seven books, each focusing on a different bodily system and each beautifully illustrated, probably by the artist Jan van Kalcar (c. 1499–1545). The books revealed in turn the structures of the skeleton, muscles and circulatory system; the distribution of the nerves; the structure and arrangement of organs in the abdominal and thoracic cavities; and the configuration of the skull and brain. The *Fabrica* was published in conjunction with an abridged companion volume, the *Epitome*, to be used as a more accessible guide for students. In line with contemporary artistic conventions, the illustrations depicted the components of the body in active, life-like poses in idealized rural settings, a feature that demonstrated the substantial skills of the anatomist and facilitated wider cultural acceptance of dissection (see Figure 7).

Although Vesalius was an ardent Galenist, his meticulous observations initiated a gradual rejection of Galen's anatomy and physiology. The *Fabrica* exposed many of the inaccuracies inherent in ancient understandings of the human body. According to Vesalius the human heart comprised four, not three, chambers and did not appear to have pores between the two ventricles, as Galen had suggested; the liver did not possess five lobes; and the

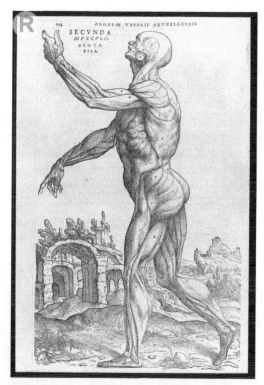

Figure 7   Andreas Vesalius, *De humani corporis fabrica*, 1543 (Wellcome Library, London)

rete mirabile, a particular network of arteries and veins that Galen understood to be the place where the *élan vital* (vital spirit) and the blood were changed into 'animal spirit', was not a feature of the human circulatory system. In subsequent editions of the *Fabrica*, Vesalius qualified his initial observations and incorporated corrections and additional details that helped to hasten the Renaissance revolution in descriptive anatomy.

Vesalius's interpretations of his anatomical findings were in turn criticized by anatomists and physicians who preferred a more literal reading of Galen's texts. Nevertheless, publication of the *Fabrica* constituted a turning point not only in Vesalius's life, as he was subsequently appointed as a court physician and military surgeon, but also in the history of medicine. After 1453, anatomy emerged as an important field of enquiry in its own right. The practice of dissecting corpses was often regarded with distaste, linked as it sometimes was to an illegal trade in dead bodies, but eventually, throughout Europe, it became the basis for a good medical education. The frontispiece to the *Fabrica* and 'The Anatomy Lesson of Dr Nicolaes Tulp' (see Figure 8), painted by the Dutch artist Rembrandt van Rijn (1606–69) in 1632, testify to the enthusiasm with which Renaissance and early modern medical students, surgeons and physicians in Padua, Bologna, Leiden, Uppsala and elsewhere crowded into newly built anatomy theatres to observe skilful dissections of the human body.

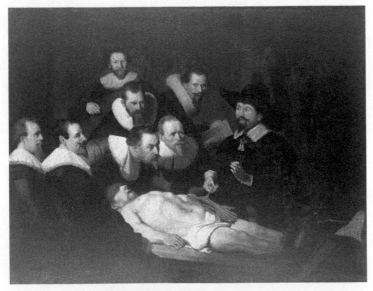

Figure 8   Rembrandt van Rijn, 'The Anatomy Lesson of Dr Nicolaes Tulp', 1632 (Wellcome Library, London)

Vesalius was by no means the first to emphasize the importance of anatomical dissection (also known as autopsy) for the practice of medicine but he was arguably the most influential. His work on the functional anatomy of the human body inspired subsequent studies by Michael Servetus (1511–53), Bartolomeo Eustachi (c. 1500–74), Gabriele Falloppio (1523–62) and Hieronymus Fabricius ab Acquapendente (c.1533–1619). Their investigations revealed many anatomical structures and their physiological importance: the Eustachian tubes connecting the throat to the middle ear; the uterine or Fallopian tubes; the presence of valves (which Fabricius referred to as 'little doors') in the veins; the mechanisms governing foetal development and birth; and the presence of the pulmonary circulation. One of the legacies of the anatomical approach was the continued reliance on clinical observation or on the use of the senses to identify anatomical structures and diagnose disease. Of course, many disputes about human physiology, such as the relative roles of men and women in conception, were not resolved by gross anatomy. Nevertheless, through the efforts of prominent physicians such as John Caius (1510–73), anatomy was also incorporated into the English learned medicine that was principally taught at Oxford and Cambridge.

Vesalian anatomy also influenced developments in surgery. In both the East and the West, surgical knowledge had traditionally been gained from the treatment of knife and gunshot wounds sustained during battle or from the management of injuries and burns suffered at home or in the workplace. During the fifteenth and sixteenth centuries, military experience continued to inform Western surgical procedures. The introduction of muzzle-loaded firearms, such as the arquebus and musket, increased the number of gunshot wounds. Musket injuries were often severe and contaminated with dirt and clothing. Surgeons dilated the wound with instruments and poured in boiling oil to cauterize it but infection made death almost inevitable. Damaged limbs were treated with similar brutality; the limb was amputated and the stump closed with a red-hot iron. In China and India, surgery was arguably less popular but it was used to remove cataracts, treat fractures and remove foreign objects from the body, and surgeons in the East were increasingly influenced by Western surgical techniques.

The most important and perhaps most controversial Renaissance surgeon was Ambroise Paré (1510–90), who practised as an assistant surgeon at the Hôtel Dieu hospital in Paris and supported the French army as a field surgeon. Drawing on the work of Vesalius as well as on the writings of Graeco-Roman and Arabic authors, Paré introduced many innovations in surgical technique. In particular, he rejected the use of hot irons and boiling oil to cauterize wounds, replacing this harsh technique with the gentler and less invasive application of dressings containing egg, oil of roses and turpentine. In addition, he pioneered the use of ligatures, or silk threads, to stem the flow of blood during amputations, improved the treatment of fractures, and provided insights into the management of childbirth. Paré is also renowned for dismissing long-standing superstitious beliefs about the healing properties of the bezoar stone, a solid mass found in the stomachs of animals. By carrying out a series of experiments demonstrating

AN ANNIVERSARY FOR PEPYS

On 26th March 1658, the London surgeon Thomas Hollier (1609–90) operated successfully on the diarist Samuel Pepys (1633–1703) to remove a bladder stone. Pepys was so grateful to have survived the surgery and to be free from pain that he celebrated the anniversary of the operation every year. 'This day it is two years since it pleased God that I was cut of the stone at Mrs. Turner's in Salisbury Court', he wrote in his diary in 1660. 'And did resolve while I live to keep it a festival, as I did the last year at my house, and for ever to have Mrs. Turner and her company with me.'

http://www.pepysdiary.com/diary/1660/03/26/

that the stone was not an effective cure for patients who had been poisoned, Paré helped to establish a more rational basis for medical practice.

Surgery was excruciatingly painful and hazardous. In the absence of reliable anaesthetics, Renaissance and early modern surgeons tied their patients to the operating table and administered herbal drinks to reduce the pain. One of the commonest and most successful operations throughout the early modern period was 'cutting for the stone' (lithotomy), which generally involved opening the bladder through the perineum and removing the stone with specially designed scalpels and forceps. Some patients survived but many procedures proved fatal. However, surgery gradually improved during the Renaissance, as the result of new anatomical knowledge and a broader commitment to humanist surgery; that is, a form of surgical expertise that combined the manual dexterity of the surgeon with the logic and learning of the physician. In this field, as in art and anatomy, ancient aesthetic standards as well as Hippocratic and Galenic wisdom initially provided the rationale for practice.

New techniques in anatomy and surgery were not routinely accepted, nor did they continue to support the authority of Greek and Roman scholars. Disputes persisted about the morality and safety of dissecting dead bodies, about the ethics of submitting patients to dangerous operations, about the professional status of surgeons, and about the relative benefits of different surgical procedures. In India and China, as well as Europe and North America, doctors disagreed about the morality of performing Caesarean sections to deliver babies, since the operation routinely led to the mother's death. In spite of ethical and practical obstacles to more elaborate surgical intervention, advances in Renaissance surgery, art and anatomy initiated the gradual rejection of Hippocratic and Galenic humoralism and promoted the emergence of fresh experimental approaches to scientific knowledge and clinical practice.

The scientific revolution

Before the Renaissance, Western understandings of human nature and the cosmos were largely Aristotelian. According to the Greek philosopher Aristotle (384–322 BCE), the universe and its components were constructed in an orderly and largely symmetrical manner. Just as the body was composed of four humours, there were also four elements, four primary qualities, four temperaments and four seasons. At all levels of life, harmony and health depended on the effective balance of these individual components. Aristotelian logic provided a comprehensive matrix for understanding the inter relationships between animals, their immediate natural environment and astrological or cosmic forces. In addition, for Aristotle and his followers, such as the Greek mathematician and astronomer Ptolemy (c. 90–168 CE), the earth was at the centre of the universe: according to what became known as the geocentric theory, the sun and planets revolved around a stationary earth.

Given the everyday experience that the earth is not moving, belief in the geocentric model of the universe was not unreasonable and it persisted relatively unchallenged for over 1,000 years. As with Galen's anatomy and physiology, objections to Aristotle's astronomy were prompted by the Renaissance revival and reassessment of classical learning. While some scholars continued to revere Greek knowledge, during the fifteenth and sixteenth centuries others began to stress the need to establish a deliberately 'modern', rather than 'ancient', science to explain the world and the cosmos. What became known in the twentieth century as the 'scientific revolution' was made possible by changing social and religious practices, by growing preoccupations with empirical evidence obtained through experimentation, and by the invention or improvement of technologies that collectively helped to reorder the world and its constituents according to new laws.

In 1543, the same year in which Vesalius published the *Fabrica*, the Prussian astronomer, mathematician and physician Nicolaus

Copernicus (1473–1543) suggested a radical reinterpretation of the universe. In *De revolutionibus orbium coelestium* (*On the Revolutions of the Heavenly Spheres*), Copernicus proposed a heliocentric model of the planets, according to which the sun, rather than the earth, occupied the central position. The apparent rotation of the sun around the earth, he argued, was simply the product of the earth's daily rotation about its own axis. Copernican astronomy was not universally accepted, largely because it contradicted long-standing Aristotelian and Ptolemaic principles and challenged religious beliefs in the divine supremacy of humanity. Nevertheless, during the seventeenth century, the heliocentric model was increasingly supported by more accurate observations of the movements of the planets and stars facilitated by the development of more powerful telescopes. In particular, the invention and refinement of the reflecting telescope by the Italian mathematician and astronomer Galileo Galilei (1564–1642) allowed scientists to study the universe in greater detail and to confirm many of Copernicus's suppositions.

As the lives of Copernicus and Galileo suggest, revolutionary approaches to the natural world were often engineered by scholars with multiple intellectual and creative talents who, like many of their late medieval Islamic predecessors, were prepared to challenge authority and extend the boundaries of human knowledge and experience. Applied to Leonardo da Vinci, Michelangelo, Galileo, Taddeo Alderotti (1215–95), Girolamo Cardano (1501–76) and others, the term 'Renaissance man' (or polymath) came to signify a capacity to contribute to learning in several different disciplines, including astronomy, natural philosophy, ethics, law, anatomy, medicine, mathematics, art and theology. The lives of these men (and they were almost all men, not women) and their impact on the world highlight the manner in which Renaissance science and medicine were part of a much wider intellectual endeavour to eradicate superstition and blind devotion to ancient authorities and to establish reliable principles of scientific enquiry. This empirical and inductive approach, expounded most clearly by the English scientist and statesman Francis Bacon (1561–1626), as well as new technologies, provided the groundwork for the subsequent discovery of the principles of planetary motion by Johannes Kepler (1571–1630) and the laws of gravity by Isaac Newton (1642–1727).

Not surprisingly, novel approaches to scientific knowledge provoked strong reactions not only from Aristotelian and Galenic scholars but also from the Catholic Church, which regarded the suggestion that the earth moved around the sun as an outrageous attack on fundamental Christian beliefs. However unreasonable it might seem to modern views, it is easy to see why Church leaders felt threatened by new scientific methods that seemed to disparage scriptural wisdom: according to Galileo, the investigation of nature should 'begin not with the Scriptures but with experiments, and demonstrations'. Several leading Renaissance figures were persecuted by religious authorities for their beliefs. Galileo's support for Copernicus provoked such strong reactions from the Pope that his work was banned and in 1632 he was summoned to Rome to stand trial for heresy. Finding him guilty, the Inquisition sentenced Galileo to imprisonment and he remained under house arrest until his death nearly ten years later. Servetus suffered an even worse fate. Despised by Catholics and Protestants alike for his objections to the belief that God existed as three persons (a doctrine known as Trinitarianism), Servetus was condemned for heresy by the Protestant reformer John Calvin (1509–64) and burned at the stake.

Acceptance of the new science was encouraged by the declining authority of the Catholic Church. Already unsettled by the disastrous spread of plague and other fatal contagious diseases in the late Middle Ages, during the sixteenth century the power of the Pope was further challenged throughout Europe by humanist Protestant reformers, who hoped to eradicate the doctrines and constitution of Catholicism. While southern Europe remained Catholic, much of northern Europe was strongly reconfigured by

Protestantism, a development that not only triggered the Thirty Years' War (1618–48), which killed thirty to forty percent of the German-speaking population, but also created novel political alliances and established the freedom to worship irrespective of denomination. Although contested and incomplete, these changes served to create a less dismissive and more tolerant approach to knowledge. At the same time, the increased migration of workers across northern Europe (initiated partly by religious conflict) necessitated new forms of poor relief and health care to deal with fluctuating levels of poverty and the spread of disease.

Developments in science were not incidental to medicine. Indeed, much of the momentum for the intellectual transitions of the Renaissance and the emergence of an empirical scientific method came from doctors who were attempting to refine Galenic anatomy and physiology and beginning to dismantle the traditional pillars of humoralism. One of the most flamboyant and controversial Renaissance figures was the self-taught Swiss physician Theophrastus Philippus Aureolus Bombastus von Hohenheim (1493–1541), who adopted the name Paracelsus to signify his superiority to the Roman medical author Celsus. Believing that knowledge came from observing nature rather than from book-learning, Paracelsus rejected the work of ancient and medieval authorities such as Galen and Ibn Sīnā. Instead of humoral accounts of health and disease, he proposed a philosophy of

PARACELSUS

My accusers complain that I have not entered the temple of knowledge through the right door. But which one is the truly legitimate door – Galen and Avicenna, or Nature? I have entered through the door of Nature. Her light, not the lamp of an apothecary's shop, has illuminated my way.

Paracelsus, *Paragranum*, c. 1530

medicine that highlighted the chemical constituents of the body. According to Paracelsus' medical chemistry, sometimes known as iatrochemistry, gout was not the product of humoral imbalance but the result of the deposition of chemicals, such as tartar, in the joints.

Paracelsian remedies were based on similar principles and shaped the treatment of diseases throughout the sixteenth and seventeenth centuries. One of Paracelsus's followers, the Flemish nobleman Joan Baptista van Helmont (1579–1644), for example, used chemical theories to displace ancient approaches to diagnosis and treatment. In addition, Paracelsus acknowledged and respected the role of traditional or folklore remedies as well as the potential for magic and religion to explain the causes and direct the management of ill-health. One of the popular approaches to disease that he adopted was the 'doctrine of signatures', a reformulation of the ancient principle of correspondence, according to which common substances were employed to treat particular organs of the body on the basis that they resembled those organs. The spotted leaves of the lungwort, for example, were thought to bear a resemblance to human lungs and were used to treat pulmonary infections. For Paracelsus and others, both the physical attributes and the medical potency of plants were thought to be God-given.

As a result of his unorthodox medical views, his commitment to political and religious reform and his conceited manner, Paracelsus attracted considerable censure from both learned scholars and political and religious authorities. Nevertheless, his work contributed to the eventual downfall of Hippocratic and Galenic theories and the formulation of more empirical accounts of the body. A belief in the primacy of direct observation was especially apparent in the discovery of the circulation of the blood by William Harvey (1578–1657). Educated initially in Cambridge and then in Padua, where he studied under Fabricius, Harvey began to practise medicine in London in 1602. He rapidly rose to fame, became a leading figure in the College of Physicians, and was appointed as royal physician to both James I and Charles I. Although Harvey was conservative and opposed to the new chemical and mechanical approaches to medicine, he was well trained in the anatomical methods that were being taught in continental European medical schools.

In 1628, Harvey published one of the most celebrated texts in the history of medicine. *De motu cordis* (*On the Motion of the Heart*) recounted the experimental evidence that he had accumulated to reveal flaws in Galen's physiology and to demonstrate the continuous circulation of the blood through the lungs and the body. Part of Harvey's anatomy depended on the dissection and vivisection of animals. However, he also conducted experiments on humans. It was during these experiments that he identified the significance of the heart in driving the flow of blood through the body and established the manner in which valves in the veins facilitated the return of blood to the heart by preventing backward flow (see Figure 9). Harvey's excitement at corroborating

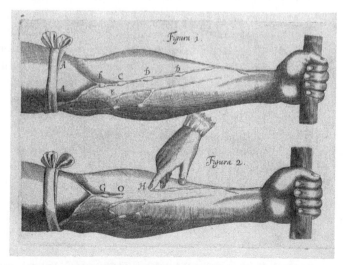

Figure 9   Illustration demonstrating the action of valves in the veins, William Harvey, *De motu cordis*, 1628 (Wellcome Library, London)

the work of Fabricius and confirming Aristotle's theory that the heart was the most important organ is evident in his written account of his discoveries: 'The heart of animals', he wrote, 'is the foundation of their life, the sovereign of everything within them, the sun of their microcosm, that upon which all growth depends, from which all power proceeds'.

Other discoveries followed Harvey's work. In the 1660s, for example, the English physician Thomas Willis described the circle of arteries that supplies blood to the brain, now known as the 'circle of Willis'. Born in Wiltshire but raised within the royalist shadows of the University of Oxford, Willis's professional fortunes fluctuated with the fall and rise of the English monarchy. He initially struggled to establish a medical practice during the 1640s and 1650s but his prospects later improved, partly as the result of the publication of a number of books on chemical medical philosophy and partly as the result of the restoration of Charles II in 1660, after which he was appointed Sedleian professor of natural philosophy at Oxford. During the following years, Willis began his intricate studies of the brain and nerves, leading to the publication of important books on cerebral anatomy and pathology and to detailed accounts of the nervous regulation of respiration, not only in health but also in clinical conditions such as asthma. Willis's emphasis on the nerves led to specific forms of treatment. Believing some forms of 'convulsive' asthma to be triggered by irritation of the muscles, nerves or brain, for example, he prescribed anti-spasmodics and sedatives such as asafoetida and laudanum to relax the airways and reduce the sense of breathlessness.

New knowledge of the body in health and sickness was made possible by the introduction of new instruments. Just as the telescope had transformed understandings of the heavens, so the microscope began to transform understandings of the body. First developed by Galileo in the 1620s, the microscope was refined and improved by Marcello Malphigi (1628–94), Antoni van

Leeuwenhoek (1632–1723) and Robert Hooke (1635–1703). Hooke's *Micrographia*, published in 1665 and illustrated by the architect Christopher Wren (1632–1723), beautifully demonstrated the capacity of the microscope to reveal the fine structures of animal and plant tissues. The increased magnification achieved by van Leeuwenhoek's modifications to the microscope allowed him to see for the first time red blood cells, bacteria and sperm cells, as well as the structure of nerves, teeth and hair. By the end of the seventeenth century, the insights provided by innovative scientific methods and instruments had radically altered anatomical and physiological accounts of the body. Reinforced by the theories of the French philosopher René Descartes (1596–1650), who regarded the body primarily as a machine, mechanical models of human form and function had begun to replace older humoral theories of health and disease.

Death and despair

What impact did Renaissance anatomy and the Scientific Revolution have on the diagnosis and treatment of disease? To what extent did patients benefit from new pathological theories? How did Renaissance and early modern doctors incorporate developments in learned medicine into their everyday practice? Did improvements in surgery alter the incidence, prognosis or patterns of disease? To what extent were European models of health and disease shaped by, or transmitted to, other medical cultures? Were Renaissance and early modern populations healthier or happier than their ancient and medieval predecessors? Although these questions are difficult to answer in precise terms, we can reconstruct, with some degree of accuracy, patterns of disease and the experiences and practices of patients, families and healers.

Between the fourteenth and seventeenth centuries, concerns about disease and death were dominated by plague. The spread of the Black Death had led to a significant decline in the European population, a phenomenon captured by a growing body of treatises and tracts charting the spread of the disease. Nevertheless, other diseases also afflicted contemporary communities. The range of conditions suffered by patients is eloquently captured by the English poet John Milton (1608–74), who himself suffered from gout and became blind. In the penultimate section of *Paradise Lost*, first published in 1667, Milton referred to the variety of physical conditions to be witnessed in early modern hospitals:

> Immediately a place
> Before his eyes appeard, sad, noysom, dark,
> A Lazar-house it seemd, wherein were laid
> Numbers of all diseas'd, all maladies
>
> Of gastly Spasm, or racking torture, qualmes
> Of heart-sick Agonie, all feavorous kinds,
> Convulsions, Epilepsies, fierce Catarrhs,
> Intestin Stone and Ulcer, Colic pangs,
> Dropsies, and Asthma's, and Joint-racking Rheums.

Although perhaps exaggerated for poetic purposes, Milton's harrowing description of the groans of pain and despair that filled hospital wards provides some insight into the common manifestations of disease and the ever-present threat of death. Many children and young people died from infectious diseases, a situation exacerbated by the malnutrition caused by periods of poverty and poor harvests. Women did not always survive childbirth. In some cases, the medicine itself caused pain and illness: 'The disease torments us on the one side', wrote the French essayist Michel de Montaigne (1533–92) in 1580, 'and the remedy on the other.' While some people, including Harvey, Milton and Pepys, lived into or beyond their sixties, more than half of the population died before the age of forty. From this perspective, little had changed since the medieval period.

Given the unrelenting spectre of death, preparations for the deathbed, burial and after-life remained important. Dying well included making a will and displaying an appropriate level of spiritual acceptance. Although many people died rapidly and in pain, the ideal death involved remaining conscious and as free from pain as possible. Funerals generally took place during the day but in the seventeenth century it became increasingly popular to be buried at night, perhaps because the simpler ceremony was cheaper. Most people were buried in unmarked graves in local churchyards but ostentatious memorials, monuments and headstones were not uncommon, particularly amongst the rich. Secular emblems of the inevitability of death, referred to as *memento mori*, were incorporated into visual art and literature in this period: a skull, or a skeleton holding a dart and hourglass were used to signify death. Although dying patients and their families might rely on medical advice and support during the later stages of illness, doctors were not usually involved at the death bed itself.

As in previous historical periods, the major causes of death were infectious diseases, patterns of which continued to be shaped by global travel. In addition to the probable import of syphilis to Europe from South America in the late fifteenth century, smallpox appears to have spread in the opposite direction during the sixteenth century. Characterized by the eruption of small pustules on the skin, smallpox was spread by direct contact or by inhaling droplets from the sneezes and coughs of infected people. The disease may well have been present in ancient Rome, as well as in India, China, Korea and Japan from at least the fourth century but it was most clearly described by medieval authors such as al-Rāzī, who wrote a treatise on how to differentiate smallpox from measles. Until the seventeenth century, the disease was rarely fatal in European communities but the introduction of smallpox into the Americas by the Spanish conquistadores in the early sixteenth century decimated the vulnerable native population. The smallpox epidemic severely weakened the Aztec Empire and its army, reducing their resistance to invading forces and facilitating the conquest of Mexico by the Spanish explorer Hernando Cortés (1485–1547).

Other infectious diseases, such as influenza, typhus, dysentery, measles and malaria, were similarly spread by military manoeuvres and religious missions or along expanding trade routes. According to some colonial authorities, the desolation of native inhabitants by imported diseases demonstrated divine justification for invasion: God, they argued, was quite literally preparing the land for white Westerners. In some cases, however, these diseases caused havoc not only amongst the colonized, conquered and exploited Indigenous people but also amongst their conquerors and oppressors. The increased use of slaves from Africa, for example, led to the spread of malaria and yellow fever to susceptible European populations and onwards across the Atlantic Ocean. Equally, immigrants to the New World were confronted by infectious diseases to which they possessed no immunity and climatic conditions to which they were not accustomed, leading to high death rates amongst settlers. Medicine played a significant part both in colonial ventures and in the continuation of slavery. While the colonial authorities relied heavily on physicians and surgeons to prevent the spread of new infectious diseases, doctors were also expected to examine slaves to guarantee their health and fitness for travel and work.

Although acute infectious diseases continued to dominate patterns of morbidity and mortality, Renaissance and early modern populations were also afflicted by chronic degenerative diseases, such as arthritis, ulcers, convulsions, diabetes and cancer. In addition, as Milton pointed out in a subsequent edition of *Paradise Lost* published in 1674, people suffered from various forms of mental illness. Along with a catalogue of organic diseases, the residents of his imaginary hospital were seized with 'Dæmoniac Phrenzie, moaping Melancholie, And Moon-struck madness'. Insanity appeared in other contemporary literary works. In *Hamlet*, for example, William Shakespeare (1564–1616) depicted both Hamlet and Ophelia as mad in different ways: while Hamlet's madness appeared to be feigned in order to reveal his uncle's murderous guilt, Ophelia demonstrated the rambling irrationality of a woman apparently driven to insanity by the death of her father and her rejection by Hamlet. Although fictional, the case of Ophelia suggests that it was commonly understood that madness could lead to suicide.

Drawing on ancient accounts of mental illness established by Hippocrates and Aretaeus, Renaissance doctors principally depicted madness in two forms. Mania was characterized by excessive and uncontrollable excitement, impulsive grandiosity and euphoria. Maniacs, wrote the Swiss physician Felix Platter (1536–1614), tended to 'express their mental impulse in a wild expression and in word and deed', sometimes to the point of obscenity and brutality. By contrast, melancholy was marked by dejection, anguish, anxiety and delusions. According to Platter and other contemporary writers, such as the London physician Philip Barrough, melancholic patients suffered from a form of 'mental alienation' or disordered thinking that led to feelings of overwhelming sadness and grief and the inability to enjoy life.

PHILIP BARROUGH ON MELANCHOLY

Melancholie is an alienation of the mind troubling reason, and waxing foolish, so that one is almost beside himselfe…The most common signes be fearfulness, sadness, hatred, and also they that be melancholious, have strange imaginations, for some think themselves brute beastes, and do counterfaite their voice and noise, some think themselves vessels of earth, or earthen pottes, and therefore they withdrawe themselves from them that they meet, least they should knocke together.

Philip Barrough, *The Methode of Phisicke*, 1583

Although still understood primarily in terms of humoral imbalance or as the product of divine displeasure, the descent into madness was thought to be triggered by a variety of personal attributes and situations. In the most prominent Renaissance or early modern work on melancholy, *The Anatomy of Melancholy*, published in 1621, the Oxford scholar Robert Burton (1577–1640) suggested that insufficient exercise, idleness, isolation, fear, shame, hatred, envy, anger, the love of gambling and excessive pleasure, pride, self-obsession, over-work, too much studying, frights, infatuations, rejections, imprisonment and poverty could all lead to mental illness. According to Burton, the cure for melancholy was to address these aspects of the patient's life. Moderate exercise of the body and mind, outdoor recreation, music and laughter, friendly company and the alleviation of poverty were all capable of reviving ailing spirits and restoring sanity.

Historians have often assumed that early modern doctors rarely examined their patients directly, relying instead on the patient's account of their symptoms to reach a diagnosis. It is certainly true that physical examination was a less prominent feature of the doctor–patient consultation than it is in the twenty-first century. Touching female patients was often precluded, due to concerns about sexual propriety and professional status. In addition, continued reliance on humoral principles rendered intimate examination of individual organs less important. Nevertheless, medical records suggest that surgeons and midwives regularly palpated their patients' abdomens and chests and performed internal examinations in the treatment of wounds, in administering physical remedies such as blood-letting and in the determination of pregnancy and the management of childbirth. Physical examination became more important with the rise of anatomical dissection, which carried with it an increased understanding of the relationship between organic pathological processes and the symptoms of disease, and with growing interest in monitoring the effects of treatment more closely.

Faced with illness and despair, Renaissance people turned to a variety of healers for support and advice. In addition to seeking the learned medicine of élite physicians, the manual and surgical skills of the surgeon, and the pharmaceutical knowledge of apothecaries or pharmacists, patients and their families consulted a range of itinerant practitioners often dismissed as quacks, charlatans or mountebanks. The choice of healer depended not only on the nature of the ailment but also on a family's financial resources and status and on the availability of particular practitioners. Although most practitioners would have been men, patients also sought the advice of women healers, including nurses and midwives. Childbirth, in particular, remained the preserve of women and some early modern midwives became acknowledged experts in the field. In 1671, Jane Sharp published *The Midwives Book*, a manual of midwifery that combined information on anatomy with advice about sexuality, conception, pregnancy, labour and the care of infants.

For many patients, local healers well-versed in domestic medicine offered the most effective information about how to prevent or treat disease. Neighbours and relatives shared information about how to prepare and apply particular ointments, food, spices and herbal remedies. In Europe, China and the Mongolian Empire, these informal traditions were sometimes recorded in printed vernacular texts and family recipe books and remained largely unchanged by scholarly developments in anatomy, medicine and science. One of the most famous early modern English collections of household cures was the recipe book compiled by Lady Johanna St John (1630–1704), who grew her herbal ingredients in a walled garden at Lydiard House, her family's country estate near Swindon.

Herbs occupied a prominent place in the management of illness. In China, the agricultural and botanical knowledge compiled by Shen-nung in the third millennium BCE remained influential. Similarly, Western doctors prescribed herbal medicines largely according to the ancient knowledge of plant-based treatments contained in the work of the Greek botanist Dioscorides, who listed approximately 700 medicinal plants (as well as many animal and mineral preparations) in his *Materia medica*. Renaissance and early modern botanists and apothecaries, such as the English herbalists John Gerard (1545–1611) and Nicholas Culpeper (1616–54), not only refined Dioscorides' accounts of medicinal herbs but also developed their own prescriptions. Culpeper was particularly keen to disseminate medical advice to the general public: one of his most popular advice books, *The English Physician*, which was first published in 1652, was reprinted more than a hundred times and is thought to have been the first medical text to be published in America. In addition to the use of medicinal herbs in food and drink, wine was also applied directly to wounds to prevent infection. Some herbal remedies (such as pine resin, stramonium and rosemary) were burned to provide inhalations for people with lung disease or to enable houses to be fumigated to cleanse the air.

Many of the herbs and medicines used by Renaissance doctors and their patients were grown and prepared locally, either as simple concoctions of individual substances or as compounds comprising two or more ingredients. Apothecaries and other local healers often cultivated plants in their own gardens and prepared distillations of spirits, herbs, sugar and spices (often known as cordials) to treat conditions of the nerves, heart, stomach and lungs and to include in perfumes, soaps and cosmetics. In many cases, botanical preparations were prescribed according to both medical and astrological principles. In recommending celandine for the treatment of syphilis, deafness, melancholy and diseases of the liver, for example, Culpeper referred both to its specific astrological characteristics and its broad curative properties. Although we now tend to regard herbalism as a form of alternative medicine, during the Renaissance and early modern periods it constituted a critical element of orthodox practice, sanctioned by the monarchy. In Britain, botanical medicines prescribed by physicians and dispensed by apothecaries were purchased and consumed by royalty and commoners alike.

New drugs were added to the Western pharmacopoeia from China, the Middle East and the New World, thereby promoting a global trade in medicine. Similarly, African herbal knowledge and healing practices (referred to as Obeah medicine in the West Indies) were carried with slaves to America. Cinchona, containing quinine and originally imported from Peru, guaiacum, sassafras, sarsaparilla, ma huang (ephedrine), ipecacuanha and tobacco were all introduced into European practice from abroad and became accepted treatments for a wide variety of conditions including fever, rheumatism, pain, asthma, kidney disease, sexually transmitted diseases, menstrual disorders, infertility and catarrh. Guaiacum was recommended to relieve the pain of rheumatism and was also included in recipes for treating colic. Sassafras was prescribed for kidney stones and tobacco was used in ointments for the treatment of wounds long before it became popular as a recreational drug.

Herbal medicine was gradually influenced by novel chemical approaches to knowledge and practice as doctors attempted to purify and standardize treatments. But there was no simple transition from the use of herbs as popular, magical remedies to their application as part of an orthodox secular medical practice. Although understandings of disease, at least amongst learned physicians, tended to focus on natural causes, belief in the role of sin, providence and divine displeasure in causing disease and death persisted. One clear example of this is the use of the 'touch' to cure scrofula, a disease that may well have been a form of tuberculosis and was characterized by enlarged lymph nodes in the neck. Known as the 'King's Evil', it was thought scrofula could be cured or alleviated by the medieval practice of the sufferer being touched by the king or by the patient touching a coin (known as an angel) that had been handled by a monarch.

Similarly, superstition often shaped attempts to ward off disease, restore health and vitality, and enhance or reduce fertility. While some couples employed herbal preparations, prayers and incantations to induce conception, others regularly used folk remedies to avoid or terminate unwanted pregnancies.

As the use of herbal remedies suggests, there were often no clear distinctions between different types or classes of practitioners, between élite and domestic medicine, or between Eastern and Western approaches to the treatment of disease, and there was no inevitable revision of medieval practices by Renaissance doctors and their patients. A wide variety of new and traditional healing practices existed side by side in a blossoming, and increasingly global, medical marketplace. However, there were substantial developments in the delivery of health care to the poor and infirm. In many cases, medicine was practised at the person's bedside. Stimulated partly by epidemics of infectious diseases and partly by more careful recording of death rates during outbreaks of plague, interest in promoting public health through state, as well as charitable, welfare services increased. Since the Southern Song period (1125–1275), the Chinese authorities had provided drugs in response to epidemic diseases and attempted to marginalize spiritual healers in order to improve the health of poor populations through state-sponsored health interventions.

Throughout Western Europe, both state and charitable forms of poor relief and welfare support were increasingly regarded as one way of alleviating the social causes of illness, such as poverty and malnutrition, and reducing vagrancy, begging and crime. One of the leading Renaissance proponents of medicine as a form of social reform was the Dutch Catholic priest and humanist scholar Desiderius Erasmus (1466–1536), whose knowledge of medicine was shaped by his extensive travels through France, England and Italy and by his attempts to alleviate his own illnesses, including dyspepsia, renal and bladder stones, and fevers. Drawing largely on humoral theories of health and disease, Erasmus emphasized the importance of diet and hygiene, explored the best methods of child-rearing in his work on marriage, highlighted the importance of maintaining ethical standards within the profession, and promoted legislative intervention to control epidemics and improve public health.

Hospitals became a key component of public health initiatives during the Renaissance and early modern period, particularly in rapidly growing towns and cities (see Figure 10). In Italy,

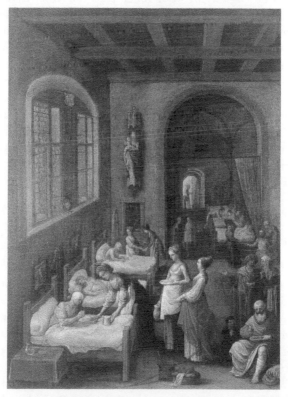

Figure 10 Saint Elizabeth offers food and drink to a patient in a hospital in Marburg, Germany, 1598 (Wellcome Library, London)

hospitals employed doctors and nurses and began to provide free medical care from the late fifteenth century. The development of poor houses and hospitals in northern Europe, as well as support for patients confined to their own homes, was facilitated by the Reformation as Protestant theologians attempted to distinguish more readily between the deserving and undeserving poor. The elderly infirm, the sick poor who could not work, and families left unsupported following the death of a husband and father would all be eligible for poor relief, paid for either by charitable donations or from the poor rates levied on householders. By contrast, those who feigned poverty or who simply refused to work were deemed unworthy of help. The deserving poor and destitute could receive 'outdoor relief', which allowed them to remain at home and work as best they could, or they could be sent to the workhouse or poorhouse, a fate that carried considerable stigma even as it provided medicine, care and prayer. In spite of the expansion of secular, rather than religious, approaches to scientific and medical knowledge, care of the soul nevertheless remained a pivotal feature of Renaissance and early modern medicine.

Continuity and change

The European Renaissance was initially marked by a return to the wisdom of classical scholars such as Aristotle, Hippocrates and Galen. Ancient knowledge shaped developments in art and anatomy and infused medical practice with humanist principles. However, paradoxically, it was the emergence of new styles of investigation and understanding that eventually served to undermine Hippocratic and Galenic accounts of health and disease and reveal flaws in ancient descriptions of the form and function of the human body. The work of Vesalius, Paracelsus, Harvey and their peers, as well as the development of new instruments such as the microscope, promoted a mechanical view of the body and encouraged the adoption of empirical investigations of anatomical structure and pathological processes. Scholarly knowledge relied increasingly on evidence, signs and experience and on more accurate measurements of patterns of disease. These developments provided the basis for the rational medicine that was subsequently developed by Western scientists and clinicians during the European Enlightenment.

Yet, in spite of these intellectual and technical transformations, it is evident that much remained the same. Between the fifteenth and seventeenth centuries, people across the world remained vulnerable to outbreaks of infectious diseases and employed similar herbal remedies and isolation to preserve or restore individual health and prevent epidemics. Although secular approaches to illness blossomed, religion continued to influence explanations of disease as well as the provision of healthcare services in both the East and the West. Charitable and state welfare initiatives offered respite and support for the sick and their families much as they had done in the late medieval period. A vibrant domestic medical culture continued to provide culinary and therapeutic advice predicated on ancient knowledge of herbs, spices and medicines and on the long-standing practices of midwifery and surgery.

Although challenged, humoralism persisted, even amongst learned practitioners. A clear example of this can be found in the work of the English physician John Floyer (1649–1734), who wrote the first treatise on asthma written in English in 1698. Floyer published widely on a range of medical techniques, including the therapeutic value of cold bathing, wrote a student textbook on the art and ethics of medicine, and invented a 'pulse watch' with a second hand for more accurate measurement of heart rate. Basing his recommendations on more than thirty years' experience of struggling to manage his asthma, Floyer dismissed the novel chemical theories advocated by Paracelsus and van Helmont. Instead, he insisted that the 'old Writers', namely Galen and Hippocrates, had discovered the most useful remedies for patients with asthma, a condition that he explained in characteristic humoral terms as a product of the accumulation of phlegm.

Floyer's approach to treatment was essentially medieval. His advice to patients focused on establishing a suitable regimen or lifestyle, supplemented by specific remedies. During the asthmatic fit, patients were to avoid fumes and smells and sit upright in an airy room to facilitate breathing. Vomiting, bleeding, blisters and cool liquors were to be administered to reduce bronchial constriction, and gentle opiates could be taken before bedtime, when symptoms often became worse. Vinegar acids, renowned for their therapeutic properties since the time of Hippocrates, were prescribed to alleviate the inflammation and suffocation of asthmatic paroxysms. Relying on the writings of Celsus, Galen, Oribasius and Ibn Sīnā, Floyer recommended a range of traditional decoctions, emetics, purgatives, digestives, sudorifics (sweat-inducers) and enemas to maintain respiratory health between attacks. One of his favourite remedies for preventing and alleviating asthma was vinegar of squills, which was prepared from the bulbs of the sea onion and had, according to Galen, originally been discovered by Pythagoras (c. 570–495 BCE).

Floyer's reliance on ancient and medieval texts and traditional remedies, combined with his aggressive dismissal of modern therapeutic fashions, suggests that the light of classical Greek, Islamic and Jewish medicine had not yet been extinguished. However, it is noticeable that the new empirical scientific methods adopted by Western doctors embraced Hippocratic, rather than Galenic, notions of disease. In recognition of his neo-Hippocratic commitments, the English physician Thomas Sydenham (1624–89) was known as the 'English Hippocrates'. Rejecting book-learning and emphasizing the importance of clinical experience, Sydenham exemplifies the gradual transition from ancient to modern forms of knowledge during the Renaissance and early modern period. Combining old and new, as well as East and West, doctors in the eighteenth century built on these foundations to provide greater hope that, with appropriate medicine, a paradise of health and happiness could be regained.

4

Hospitals and hope: the Enlightenment

In the autumn of 1787, above 600 children were inoculated for the small-pox at Geneva, who all, except two, happily recovered from the disease. As the natural small-pox is fatal to one in five, this general inoculation may fairly be computed to have saved 118 lives.

The Times, 25 April 1788

The eighteenth century was an age of confidence, ambition and hope. The world became smaller, as people travelled more easily across borders and seas. Social mobility increased, as the creation of a global marketplace generated greater opportunities for fame and fortune and the boundaries between classes, at least in terms of economic status, became more porous. Rising levels of literacy, dramatic increases in population, and the cataclysmic processes of urbanization and industrialization promoted the emergence of a modern consumer society. New commodities became available and novel forms of authority began to challenge traditional patterns of knowledge and power. Science and medicine marginalized older alchemical and metaphysical visions of the cosmos and religion was increasingly rejected in favour of secular accounts of liberty and social order. Reason replaced revelation both as a source of understanding the natural world and as a means of ensuring technological innovation and the expansion of Western civilization.

These momentous social, cultural and intellectual transitions constitute what both contemporaries and historians have referred to as the Enlightenment. Known in Germany as *Aufklärung* and in France as *siècle des Lumières*, the Enlightenment was a European-wide movement that occurred between the late seventeenth century and the French Revolution in 1789. The precise form and impact of Enlightenment ideology varied according to region but similar trends can be identified in England, Scotland, France, Germany, Holland and Russia. The growing optimism of Enlightenment figures was not entirely misplaced. Scientific approaches to the natural world, and particularly to the human body, were beginning to transform patterns of health and the treatment of illness: during the eighteenth century, remedies for smallpox and scurvy reduced mortality rates from these diseases and surgery became a more acceptable and more effective option for the treatment of wounds, injuries and some internal conditions.

Belief in progress was not universal. Hope was tempered by scepticism and by a growing awareness of the dark side of modern civilization, a phenomenon that historians have referred to as 'the pathology of progress'. According to some Enlightenment authors, increasing luxury and over-indulgence, promoted by both individual greed and deepening cultural commitments to consumption, served only to increase the prevalence of chronic diseases such as gout, mental illness, rheumatism and obesity. The relentless pursuit of wealth, they argued, was jeopardizing health. Population growth, urbanization and the expansion of settlement frontiers in America, Europe and China were blamed for rising levels of poverty, the persistence and spread of infectious disease such as plague, and the high infant mortality rates that were recorded in many cities. Increased trade and commerce paved the way for competition between merchants and nations and for strident professional disputes between various classes of healers and carers. Enlightenment and civilization, insisted the discontents, had brought with them disease, disorder and despair as well as the prospect of progress and cure.

Nowhere is the double-edged nature of the Enlightenment more visible than in the life of the Scottish physician George Cheyne (1671–1743). Born in Aberdeenshire, Cheyne was an advocate of the new mechanistic philosophy of the world evident in the physics of Isaac Newton and the mechanical model of disease promoted by Archibald Pitcairne (1652–1713). According to Cheyne, the body was a machine, the health of which depended not on the balance of humours but on the effective working of its solid, physical constituents, the bones, muscles, tissues and organs. In particular, Cheyne insisted, the state of the nerves was crucial for health: weak nervous tone, caused by what Cheyne referred to in 1733 as 'continued Luxury and Laziness', would lead to chronic disease and pain. While Cheyne remained confident about the capacity of modern medicine to prevent or treat mental and physical illnesses, his own experiences belied his optimism. Throughout his life, Cheyne struggled to cope with gout, melancholy and obesity generated by an immoderate lifestyle: at one point in his life, his weight ballooned to 32 stone (203 kilograms). Many of Cheyne's books, which became immediately popular, recommended a regimen that had periodically helped him to recover his own health: a diet of milk and vegetables combined with air, exercise and appropriate mild medication. For Cheyne, as for many Enlightenment authors, the secret to health and happiness lay in avoiding the perils of civilization and returning to a more simple, rational and natural way of life.

To what extent did Enlightenment medicine alter the treatment and patterns of disease? Did the theories of Cheyne and his contemporaries signal a departure from, or a continuation of, medieval and early modern accounts of health and disease? How did experiences of the body and mind in health and disease change during the eighteenth century? Were Western Enlightenment models of disease shared or shaped by other cultures? What were the social, political and cultural determinants of Enlightenment medicine? More fundamentally, was there a medical

Enlightenment? There are no simple answers to these questions. But by exploring contemporary understandings and treatments of disease; the growth of a marketplace for medical services; the rise of general and specialist hospitals; the treatment of insanity; and the place of women in the management of pregnancy, labour and childcare, we can begin to see Enlightenment medicine as an amalgam of tradition and innovation, an intriguing blend of hope and despair.

Theories and therapies

Medicine was a key component of the Enlightenment mission to emancipate the world from fear, poverty, suffering and death. Modern rational medicine bolstered a common quest for health when sickness and pain constituted a persistent menace. Keeping well was as much an enduring preoccupation for eighteenth-century people as it had been for their ancestors. 'Health', wrote the English author Samuel Johnson (1709–84) in 1782, 'is the basis of all happiness this world gives'.

Yet attitudes to, and experiences of, disease were not consistent. For Johnson, disease was an inevitable accompaniment to life; for others, such as the English physician and natural philosopher Erasmus Darwin (1731–1802), ill health was more likely to be the product of external, and often malevolent, circumstances. In addition, disease clearly varied in frequency and type according to age, sex and lifestyle and was still understood in terms of providence and sin, as well as in terms of material changes in environment, climate, diet and behaviour.

Eighteenth-century communities suffered from a range of conditions. Epidemic disease remained a prominent cause of ill health and death, particularly amongst the very young, the elderly and the malnourished. Although plague began to recede in Europe after a major epidemic in Marseille in the 1720s, other infectious diseases continued to decimate populations around the world. Smallpox was a potent cause of scarring, blindness, impotence and, in vulnerable populations, death. Children living in cities were particularly at risk. Measles, syphilis and tuberculosis were endemic. Present in Europe, Africa and Asia for many centuries, influenza increasingly appeared in the form of major epidemics and pandemics during the seventeenth and eighteenth centuries. Dysentery and diphtheria caused fevers, diarrhoea and dehydration and contributed to high infant mortality rates: across Europe, approximately fifty percent of children died before the age of ten. The spread of such infectious diseases was facilitated by commerce across national borders and between urban centres, and by crowded living conditions and poor diets.

While contagious diseases dominated patterns of sickness, chronic conditions also challenged the health and happiness of eighteenth-century people. Particularly in Europe and North America, nervousness, heart disease and obesity were thought to be on the rise, a phenomenon precipitated by individual gluttony as well as by the collective fashion for luxury and intemperance amongst the prosperous classes. Although Western populations in the early twenty-first century believe rising levels of obesity to be a recent phenomenon, being overweight, or corpulent, was not unusual in the eighteenth century. Obesity was generally understood to be the result of laziness or sin but some medical writers insisted that a tendency to put on weight could be inherited. Perhaps the most famous obese person in history was the English gaoler Daniel Lambert (1770–1809), whose body became a spectacle for visitors to London and who was immortalized in a wax model that was exhibited in America. When he died at the age of 39, Lambert weighed nearly 53 stone (336 kilograms) and it took twenty men to roll his coffin into the grave.

George Cheyne was not alone in admonishing his compatriots for living a life of harmful extravagance. In 1771, the British physician William Cadogan (1711–97) similarly insisted that gout, which had been recognized by ancient Greek authors and was sometimes referred to as *podagra*, was not inherited, as many people believed, but the result of 'Indolence, Intemperance and Vexation'. From Cadogan's perspective, the treatment of gout was self-evident: 'the remedies are obvious', he wrote in *A Dissertation on the Gout*: 'Activity, Temperance and Peace of Mind'. Other clinicians, such as the American polymath Benjamin Franklin (1706–90), echoed Cadogan's prescription for wholesome living. Although Franklin himself struggled with obesity, weighing over 21 stone (133 kilograms) when he died, he was an ardent advocate of swimming as a particularly healthy form of exercise.

Nor was Cheyne the only prominent Enlightenment figure to suffer from his excesses or to write about his illnesses. Thomas Sydenham, Samuel Johnson and the English historian and Member of Parliament Edward Gibbon (1737–94) all bemoaned the impact of gout on their lives. Yet, perhaps paradoxically in retrospect, gout constituted a badge of honour, a desirable condition that was not only thought to protect against more serious illnesses but also signified wealth, rank and personality. Considered to be more common amongst men than women, and amongst the inhabitants of civilized Western nations than their Eastern counterparts, gout supposedly revealed the constitutional superiority of the educated élite. Of course, this reading of chronic disease as a mark of social and intellectual dominance was not routinely accepted. The intemperance, self-importance and hypochondria of the upper classes were often ridiculed by contemporary novelists, playwrights and cartoonists (see Figure 11).

Although humoralism persisted as a framework for explaining ill-health, particularly in popular advice books, it was increasingly undermined by an Enlightenment focus on careful observation, classification and quantification. Like the natural world, the body became an object of rational enquiry, open to physical examination both in life and death. Much diagnosis still relied on listening to the patient's narrative of their condition but inspection and palpation of the surface of the body, as well as occasional internal

Figure 11   Etching of an obese, gouty man by James Gillray, 1799 (Wellcome Library, London)

examinations, were increasingly used to support diagnosis in spite of the constraints imposed by standards of modesty and professional etiquette. Inspection of the tongue and eyes, measurement of the pulse, and palpation of the abdomen were used to identify the presence and severity of disease and to categorize conditions more accurately in both Western and Eastern medical traditions. The interest of European doctors in examining and measuring bodily structures and activities in sickness and health was promoted by growing awareness of Chinese techniques. The pulse watch, introduced by Floyer in the early eighteenth century, was modelled directly on Chinese approaches to detecting the speed, rhythm and quality of the pulse, which had been an integral and evolving feature of Chinese diagnosis for many centuries.

In 1761, the Viennese physician Leopold Auenbrugger (1722–1809) introduced the practice of percussion. After watching his father tap barrels of beer to determine how much fluid was left in them, Auenbrugger suggested that a similar procedure could be used to demonstrate the accumulation of fluid in the chest and abdomen or the enlargement of organs. Percussion was not regularly employed by doctors until Auenbrugger's work was rediscovered by the French physician Jean-Nicolas Corvisart (1755–1821) in the early nineteenth century, but the approach testifies to a growing Enlightenment commitment to understanding disease in terms of solid lesions rather than humours and to contemporary interest in empirical observation and examination as the basis for the effective practice of medicine.

New theories of disease reinforced the value of these diagnostic procedures. Drawing on the writings of Descartes and the French physician Julien Offroy de La Mettrie (1709–51), the proponents of mechanical and chemical models of the body highlighted the ways in which bodies behaved much like machines. Physiological processes, such as the circulation of the blood or the digestion of food, were achieved by the co-ordinated function of organs operating according to physical laws. God was not excluded from these accounts of human health and disease but was consigned to the status of an intelligent designer. As part of these mechanical schemes of bodily function, many doctors emphasized the importance of the nervous system in determining the appearance and natural history of disease. Both acute and chronic diseases were thought to be generated or exacerbated by insufficient or excessive nervous tone, leading to diminished or exaggerated sensitivity and irritability of the organs. For this reason, doctors prescribed either stimulants or relaxants to restore normal tone to a depleted or overactive nervous system.

Accounts of health and sickness in terms of specific organs rather than humours were reinforced by the expansion of anatomical dissection, which was increasingly taught in private anatomy schools. The practice of autopsy was largely shaped by continental European physiologists and pathologists during the late eighteenth and early nineteenth centuries. Interest in locating the anatomical sites of various diseases was evident in the work of Giovanni Battista Morgagni (1682–1771) and his Italian peers, who attempted to correlate clinical symptoms with post-mortem findings. Stimulated partly by Morgagni's work, which was published in Latin in 1761 and translated into English in 1769, the quest to determine the origins of diseases at the level of organs and tissues spread rapidly. In Britain, Matthew Baillie (1761–1823) employed post-mortem evidence to provide clear descriptions of the pathology of cirrhosis, emphysema and pneumonia. His major work, *Morbid Anatomy of Some of the Most Important Parts of the Human Body*, first published in 1793, was translated into many European languages and appeared in several English and American editions. The finest and most influential exponent of the new localized pathology, however, was the French surgeon, Marie François Xavier Bichat (1771–1802). Although he accepted that some conditions, such as nervous diseases and fevers, were not addressed adequately in his system of investigation, Bichat's writings provided an innovative conceptual and practical framework for exploring the anatomical and histological basis of disease.

Enlightenment physicians gradually embraced the new approach to pathology. According to the Scottish doctor William Cullen (1710–90), who served as president of both the Royal College of Surgeons of Glasgow and the Royal College of Physicians of Edinburgh, the dissection of cadavers was 'one of the best means of improving us in the distinction of diseases'. In addition, evidence from the examination of dead bodies proved increasingly useful in the courts. In cases where an unmarried mother was suspected of concealing her pregnancy to avoid her shame and subsequently accused of murdering the infant at birth, careful examination of the child's body often revealed evidence that the child had either been born dead or died naturally, thereby saving the mother from conviction and execution. However, it is

THE 'LUNG TEST'

One of the principal ways of determining the cause of death in cases of suspected newborn child murder was the 'lung test', which had been introduced in European courts in the seventeenth century. This involved dissecting the infant's body and putting its lungs in water. If the lungs 'swam' or floated, the child was presumed to have been born alive and murdered; if they sank, the courts assumed that the child had been stillborn and the mother was acquitted.

evident that the new approaches to anatomy that were developed around the end of the eighteenth century did not entirely replace contemporary classifications of disease on the basis of individual symptoms and observable signs. Indeed, many clinical theories merged ancient and modern forms of knowledge of the body in sickness and in health.

The leading medical theorists of the eighteenth century did not come primarily from Italy, as they had done during the Renaissance, but increasingly from the expanding universities of northern Europe, most notably in Germany, England, Scotland, Holland and France, or from North America. Physicians such as Friedrich Hoffmann (1660–1742), Hermann Boerhaave (1668–1738), Georg Stahl (1659–1734), Albrecht von Haller (1708–77) and the American Benjamin Rush (1746–1813) developed and promoted their own systems of medicine. Although they often shared a commitment to explaining disease in terms of local pathology, they nevertheless differed in focus and interpretation. One of the features of Stahl's work, for example, was his insistence that bodily functions could not be entirely reduced to mechanical laws. Stahl believed that human behaviour was also regulated by a soul or 'anima' that served as the seat of consciousness and the source of vitality and resistance against disease.

Although Enlightenment physicians emphasized the mechanical determinants of health and disease, echoes of humoralism persisted. Leeches had been employed by many ancient medical cultures to remove blood and restore humoral balance and they continued to be used during the medieval and early modern periods. Once applied to the skin, leeches sucked blood until they had gorged themselves and then fell off. Towards the end of the eighteenth century, medicinal leeches became an important commodity traded across European borders. British and Irish doctors regularly imported leeches from France and Germany to treat infectious diseases such as cholera and typhoid, to remove blood from women suffering from puerperal fever, or to relieve the pain and inflammation associated with ear infections. Leeching became so popular that businesses were established solely to import and supply leeches to chemists and hospitals. The use of leeches gradually declined during the nineteenth century but it did not entirely disappear. Indeed, from the 1980s the application of leeches became fashionable again, as a means of reducing swelling and promoting the healing of wounds.

It would be a mistake to focus on élite Western science and medicine as the only sources of knowledge during the eighteenth century. Eastern physicians, such as the Vietnamese doctor Lán Ông (1720–91), developed approaches to epidemics and the diseases of old age that drew heavily on clinical observation and awareness of the detrimental effects of rural poverty. Ông particularly emphasized the role of diet, hygiene, housing, clothing and education in promoting health and compiled an influential study of native medicinal plants. There was also considerable cultural exchange between Eastern and Western traditions. European doctors and their patients benefited substantially from the import of medicines and healing practices such as acupuncture from China, Japan, India, Africa and the American colonies, and the increasing consumption of exotic spices and drugs, such as cinchona bark for the treatment of fevers or sarsaparilla for gout and joint pains, created considerable profits for traders. Although Western doctors increasingly regarded Chinese medicine as inferior, because its theoretical basis appeared to be unsubstantiated by scientific evidence and because Chinese practitioners tended to devalue the patient's narrative when making a diagnosis, there were similarities between Western and Eastern approaches to diagnosis and treatment, particularly in relation to the use of certain herbs.

Medical knowledge and healing practices travelled in other directions, particularly through the medium of an expanding slave trade. In both North and South America, the indigenous populations learned much from slaves and from African practitioners of shamanism and magic. Similarly, medicine in the East was shaped by the transfer of Western knowledge through missionaries and their doctors. Proponents of Islamic, Chinese, Japanese and āyurvedic medicine borrowed elements of European practices, particularly in relation to anatomy, surgery and the study of herbal remedies. In some cases, however, strident anti-colonial campaigns encouraged local populations to resist the incursion of Western medicine and defend their medical traditions.

The dominance of élite Western medicine was also limited by the continuation of a vibrant domestic tradition, transmitted by herbal textbooks (or pharmacopoeias) and domestic recipe books. Families and communities often exchanged animal and herbal remedies, as well as culinary preparations, for common disorders: the skins of dead puppies were used to prepare skin ointment; snail juice and hen's dung were thought to cure eye conditions; and musk, camphor, castor, asafoetida and flowers of zinc (containing zinc oxide) were recommended for the treatment of asthma. New remedies were added to this armamentarium by local practitioners. In 1785, the English physician William Withering (1741–99) introduced the use of foxglove for the treatment of dropsy (a condition characterized by the accumulation of excess fluid in the body) after discovering a family recipe invented by 'an old Woman in Shropshire'. Withering identified digitalis as the active ingredient of foxglove and the drug was regularly prescribed not only for dropsy but also for tuberculosis and epilepsy. Although side-effects initially limited its use, interest in the pharmacological potency of digitalis persisted and it was subsequently used to treat conditions such as atrial fibrillation and heart failure.

To what extent did novel medical theories and practices affect the health and survival of patients during the eighteenth century? One area where patients benefitted was in the surgical treatment of wounds, fractures and other localized conditions. Although there were no reliable anaesthetics, other than alcohol and opium, greater knowledge of anatomy encouraged improvements in surgical expertise. As in previous centuries, injuries sustained during battle allowed army and navy surgeons to refine their surgical skills and to apply them to civilian practice once they left the armed forces. In France, Napoleon's chief military surgeon, Dominique Jean Larrey (1766–1842), was responsible for introducing stretcher-bearers, mobile field hospitals and carriages or ambulances to allow surgeons to transport and treat wounded soldiers more rapidly. Across Europe and North America, military surgeons became involved in inspecting recruits to ensure that they were physically and mentally fit to fight.

Although surgeons usually treated minor complaints, such as boils, ruptures, ulcers and rotten teeth, they performed invasive and often brutal surgery that was generally destined to end in fatality. Cataracts were removed, injured limbs were amputated, hernias were repaired and cancerous growths were excised. In spite of technical improvements, surgery remained perilous. Emergency Caesarean sections were carried out to save the lives of unborn infants, although the operation routinely ended in the death of the mother. The pain and trauma of surgery without anaesthesia in this period was hauntingly captured by the English novelist Fanny Burney (1752–1840), who had a breast tumour removed in France in 1811. 'Yet – when the dreadful steel was plunged into the breast – cutting through veins – arteries – flesh – nerves – I needed no injunctions not to restrain my cries', she wrote in a letter to her sister Esther. 'I began a scream that lasted unintermittingly during the whole time of the incision – & I almost marvel that it rings not in my ears still! so excruciating was the agony.'

The most impressive operation remained lithotomy, the removal of a bladder stone. William Cheselden (1688–1752), surgeon at St Thomas's Hospital in London, was renowned for his speed and dexterity: his skill in performing the 'lateral operation' that had been devised by an itinerant French practitioner, Jacques de Beaulieu or Frère Jacques (1651–1719), allowed Cheselden to remove a stone in less than a minute. The leading British surgeon of the eighteenth century, however, was probably John Hunter (1728–93), who wrote extensively on the surgical treatment of gunshot wounds, shock and inflammation and who successfully removed a large tumour from a patient's parotid gland in 1785. Together with his older brother William (1718–83), who opened his own anatomy school and museum at a purpose-built house in Great Windmill Street in London during the 1740s, John contributed to the growing reputation of surgeons during the Enlightenment. In Britain, the rising status of surgery was evident in the formal separation of surgeons from barbers in 1745, when the Company of Surgeons was formed, and subsequently in the establishment of the Royal College of Surgeons in 1800.

Another key development that influenced the lives of patients in the eighteenth century was the spread of inoculation for smallpox, which involved the application of infected material to trigger a protective immune response and prevent more serious infection. Endemic in Asia and Africa, the most virulent form of smallpox carried a high mortality and spread globally via trade routes. Less fatal forms were more common in Europe but deaths were not unknown and even those who survived were often severely scarred with the characteristic pockmarks of the disease.

In 1694, Queen Mary II (1662–94) contracted smallpox and died at Kensington Palace in December of that year. Aware of the dangers of contagion, Mary had forbidden her family to visit her while she lay dying, to protect them from the disease. As Mary's death suggests, there were no effective means of preventing or treating smallpox in the West and whole families could lose their lives to the disease.

For many centuries, Turkish, Chinese, Arabic, African and Indian healers had been protecting patients from smallpox by deliberately infecting, or inoculating, them with a small quantity of matter from a smallpox pustule, a procedure known as variolation, named after the Latin term for the disease, *variola*. Although some patients died from this procedure, many contracted only a minor form of the disease and became immune to further fatal infection. Variolation was introduced into western Europe by Lady Mary Wortley Montagu (1689–1762), who had observed the technique during a trip to Turkey in 1717. Lady Montagu had herself been scarred by smallpox and had her children inoculated to prevent them contracting the disease. When she returned to England, her vigorous promotion of the method led to its widespread adoption not only in Europe but also in America, where the procedure was introduced by the Puritan minister Cotton Mather (1663–1728) and first used in 1721. Although there was opposition to variolation, partly because it was a foreign practice and partly because of fears about the risks involved, leading physicians began to support its use after demonstrating its safety on criminals and orphans. Variolation also received royal approval. In Britain the Prince of Wales, later King George II (1683–1760), had his children inoculated and in Russia the Empress Catherine the Great (1729–96) and her son were successfully inoculated. Popular demand for variolation increased and the technique was applied by Poor Law doctors as well as by doctors working on their own or in small private nursing homes.

INSCRIPTION FROM A MONUMENT IN THE
NORTH NAVE OF EXETER CATHEDRAL

Here lies Sarah Lee wife of
John Lee of the Hospital of Lyons
gentleman, daughter of Nicolas
Rooke, Rector of Dartington in the County
of Devon who died
on December 2nd AD 1713
aged 23
Here also, with her seven months old daughter
is buried Rebecca
daughter of Charles Heron
of this city gentleman
and second wife of the aforesaid John Lee
who died on May 23rd 1718, aged 28
Each was devout, virtuous, modest and remarkable
not only for her kindness to all
but also for her love for her husband
Equipped with equal virtues
each was seized with an equally bad attack of small-pox
a disease most ruthless
to pregnant women
they were carried off

In spite of variolation, smallpox continued to ravage communities. Frequent epidemics led to high mortality rates across Europe throughout the eighteenth century. The scourge of smallpox was more effectively and more safely combated by the introduction of vaccination, a technique that involved infecting patients not with smallpox but with pus from the closely related cowpox or *vaccina*, derived from the Latin term for cow. In the 1770s, Edward Jenner (1749–1823) began to investigate popular beliefs that infection with cowpox protected farm workers against smallpox. In 1796, Jenner tested the theory by infecting a young boy, James Phipps, with fluid taken from a cowpox pustule on the wrist of a dairymaid. Two months later, Jenner demonstrated Phipps's resistance to smallpox. Jenner published the results of further experiments in 1798 and his work was rapidly translated into most European languages. Although it was criticized by 'inoculators', because it threatened their profits, vaccination spread worldwide, helping to reduce morbidity and mortality from a major epidemic disease.

Other successful remedies were introduced during the eighteenth century. In a series of experiments carried out in the 1740s and 1750s, the Scottish physician James Lind (1716–94) demonstrated that the consumption of citrus fruits could protect sailors against scurvy, a condition that caused bleeding gums, swollen joints, heart failure and death. Lind's treatise on scurvy was first published in English in 1753 and subsequently translated into French, Italian and German. While away at sea on HMS *Salisbury*, a British naval vessel, Lind tested a variety of treatments on twelve sailors with scurvy. 'The most sudden and visible good effects', he reported in his treatise, 'were perceived from the use of oranges and lemons'. Although Lind may not have been the first ship's doctor to advocate citrus as a cure, his work proved influential. By the end of the eighteenth century, the regular provision of lemons, limes and sauerkraut to sailors ensured that scurvy was largely eradicated from the British navy.

The rise of surgery, the success of variolation and vaccination and the decline of scurvy reinforced Enlightenment optimism about the capacity of medicine to combat disease and improve health. However, innovative approaches to treatment did not replace more traditional forms of health care either in the home or the hospital. Nor did private practice preclude greater state involvement in attempting to improve the health and fitness of poorer sections of the population. Indeed, during the last half of the eighteenth century in particular, medical science, charitable concerns about public health, and the rising consumption of medicines generated new sites for health care and encouraged the expansion of a thriving and competitive trade in health-care services.

The medical marketplace

In both the East and the West, sick patients and their families continued to consult a wide variety of medical practitioners depending on their clinical condition, their status and means, their location and their religion. Healers varied considerably in qualifications, expertise and the prices they charged for their services. In Europe, physicians continued to dominate the profession in terms of status and money but surgeons and apothecaries were a regular feature of the medical landscape. Unlicensed itinerant practitioners, sometimes dismissed by their licensed competitors as quacks, charlatans or mountebanks, offered services to those unable to afford more elaborate or fashionable interventions. Rising affluence, the increased availability of imported medicines, and a growing belief in the power of medicine to stave off disease and death encouraged the increased sale and consumption of pills, potions and nostrums. Evidence from records of household finances suggests that, as the eighteenth century progressed, families spent a greater proportion of their income on medical services, particularly during periods of severe illness or as they approached death.

As many historians have suggested, the range of healers and remedies available, along with an increased power to purchase, created a medical marketplace in which practitioners were entrepreneurs, competing for patients and patronage. As a consumer, the patient was in control of the clinical consultation, which often still took place in the patient's home rather than in a hospital or clinic. Patients chose their healers not only on the basis of the theories and treatments that they offered but also according to their personal characteristics and professional status. Given the limitations on the efficacy of many treatments, a practitioner's ability to communicate and empathize with patients and families may have been more important than his (or rarely her) capacity to cure.

In an attempt to appeal to patients and to create a monopoly on practice, physicians, surgeons and apothecaries frequently lamented the manner in which ignorant and illiterate quacks exploited patients. Similar complaints were levelled at irregular practitioners by the wider public. According to the English dramatist Ben Jonson (1572–1637), the quack was nothing more than a 'lousy fartical rogue'. In reality, however, the boundaries between licensed and unlicensed practitioners were blurred at least until the late eighteenth and early nineteenth centuries. The remedies offered by different practitioners were often similar and regular doctors were satirized as often as their irregular peers, for their arrogance and empty rhetoric, and for the risks associated with heroic treatments.

During the course of the eighteenth century, changes in medical education, the creation of medical societies, the rise of medical journalism, the growth of specialist institutions, and greater regulation of the profession led to clearer differentiation between practitioners. While Western physicians continued to require a university education, the training of surgeons became gradually more formal, shifting surgery from a craft to a profession. Surgeons were expected to have taken courses in anatomy and to have gained seven years of experience as an apprentice before passing an increasingly rigorous oral examination. After 1811, British surgeons were formally required to attend at least one anatomy and one surgery course and from 1813 they were expected to have gained one year's experience in a hospital. The cost of apprenticeship not only earned many surgeons considerable additional income but also led medical students to work in, and eventually inherit, the family practice. Throughout early modern Europe it was also possible for the widow of a surgeon (like the widows of other master craftsmen) to continue her husband's business.

Apothecaries learned through a similar apprenticeship system and there was no formal teaching from the Society of Apothecaries of London until 1803. Prospective apothecaries were sometimes apprenticed as young as thirteen years old, working in the apothecary's shop or visiting patients at home. In light of the difficulties of earning a living in the crowded medical marketplace of bustling urban centres or in isolated rural locations, many doctors were dually qualified, as surgeon-apothecaries, which allowed them to offer their fee-paying patients a wider range of services. Many of these practitioners treated minor illnesses and injuries, prescribed and dispensed medication and delivered babies. Surgeon-apothecaries were the earliest general practitioners and became key participants in nineteenth-century plans to reform the profession, not only in Britain but also in continental Europe. In 1815, the passage of the Apothecaries Act led to a restructuring and formalization of the education of British apothecaries, partly in response to the rapid expansion of druggists and chemists, who were competing with apothecaries for the right to dispense medicines.

The improved education of Western physicians, surgeons and apothecaries was facilitated by the formation of provincial medical societies, the publication of specialist journals and the expansion of hospital-based training. Like their modern counterparts, Enlightenment doctors were concerned about the wealth of new clinical and scientific knowledge that they needed to assimilate into their practice. Publications such as the *Medical Transactions of the Royal College of Physicians*, first published in London in 1768, and the quarterly *Medical and Philosophical Commentaries*, which first appeared in Edinburgh in 1773, were intended to provide overworked practitioners with a summary of recent discoveries and medical cases and observations. In addition to disseminating new information, societies and journals also generated a greater sense of professional identity amongst physicians, surgeons and apothecaries.

From the mid-eighteenth century, the training of European physicians and surgeons increasingly took place in hospitals, where apprentices paid to attend ward rounds. By 1750, all seven London hospitals accepted surgery students; by 1780 four also took trainee physicians and five took apothecary students. Excluded from Oxford and Cambridge, religious dissenters from the Church of England, such as the Quakers, Anabaptists and Puritans, often travelled either to continental European universities as part of what could be regarded as a medical Grand Tour, or to Edinburgh and Glasgow, where teaching included both lectures and bedside instruction. In Scotland, one of the principal sites of the European intellectual, scientific and medical Enlightenment, the first university professors of medical science were appointed in the late eighteenth century. These developments in education and training had an impact beyond Europe. During the eighteenth century, Jesuit missionaries carried their learning to China, where they introduced new surgical techniques, novel medicines and vaccination, and from the early nineteenth century began to establish missionary hospitals and medical schools organized along European lines.

European hospitals were diverse in nature, structure and financial status. The older medieval hospitals, such as Bart's and St Thomas's in London, tended to continue as charitable institutions, catering for the deserving poor and offering doctors opportunities to practise medicine, instruct students, and make connections with rich hospital benefactors who might then become lucrative private patients. These were largely general hospitals, treating a variety of surgical and medical cases. In some countries, such as Spain, France and Germany, state involvement in hospitals was more prominent, not only in terms of providing care and support for the poor but also in providing medical services to the rest of the population. However, anxieties about the capabilities of state-funded doctors and the professional capacities of nurses, who were often criticized for being ignorant and negligent, led to calls for greater regulation.

New hospitals were built across Europe during the second half of the eighteenth century. These institutions were established and sustained by religious and charitable principles, military necessity and municipal concerns about public health. Many were specialist institutions for distinct groups of patients. 'Lock hospitals' were built in Britain and its colonies to control and treat venereal diseases, such as syphilis and gonorrhoea, and particularly to prevent the spread of disease among soldiers and sailors. Maternity hospitals, such as the Lying-In Hospital for Married Women in London (1749) or the Lying-In Hospital of the City of New York (1798), were established to improve the management of women during pregnancy, labour and childbirth. And hospitals such as the London Dispensary for Curing Diseases of the Eye and Ear (later Moorfields Eye Hospital), founded in 1805, were built to treat patients with particular conditions. In addition, patients could obtain medical advice and treatment from dispensaries, which operated very much like modern outpatient clinics. Poor, elderly and infirm patients might also be cared for in wards or infirmaries attached to workhouses, which were supported financially by local ratepayers.

The proliferation of hospitals in the eighteenth century also owed much to a burgeoning interest in charity and philanthropy, particularly amongst the emerging middle classes in many European countries. The growth of commerce and financial opportunity created a generation of *nouveau riche* entrepreneurs, who sought innovative ways of spending their new-found wealth. Donating to philanthropic ventures offered an attractive means of advertising their commercial success, demonstrating their humanistic principles and establishing their superior social status. In Britain, charitable reformers were driven not only by a humanitarian commitment to treating the poor and needy with compassion but also by a belief in the importance of maximizing the utility of the population in order to improve the wealth and imperial strength of the nation. Within this ideological context, sometimes referred to as 'mercantile humanism', medicine became a major focus for charity.

Philanthropic initiatives promoted by British merchants included the establishment of the Magdalen Hospital for the Reception of Penitent Prostitutes, founded in London in 1758 to provide prostitutes with an opportunity to escape their 'wretched and distressful circumstances'. One of the leading supporters, and first chairman, of the Hospital was Jonas Hanway (1712–86), who also campaigned for improvements in the working conditions of young chimney sweeps and who worked for the Russia Company, an organization set up by London merchants in the mid-sixteenth century to boost European trade. Of course, homes for prostitutes were not new: during the Middle Ages, several Magdalen homes had been founded across Europe as Catholic institutions designed to rescue and reform 'fallen women'. During the eighteenth century, however, these homes were increasingly regarded not only in religious and moral terms but also in relation to their potential to remove such women from the streets and return them to health and respectable employment.

Perhaps a more striking example of Enlightenment 'mercantile humanism' is the London Foundling Hospital, which was first conceived of by the sea captain Thomas Coram (1668–1751) in the late 1730s and opened its doors to unwanted children in 1741. Coram had been shocked by the number of neglected, homeless infants and children living and dying on the streets of London, a state of affairs attributed partly to the shame of those single women who abandoned their children rather than face punishment for having an illegitimate child, and partly to the apparent frequency with which mothers were ruined by the availability and consumption of cheap liquor, especially gin (see Figure 12). The national benefits of the Foundling Hospital, like those of similar institutions across Europe, were thought to be two-fold. In the first place, it offered single mothers an opportunity to

Figure 12   'Gin Lane', an engraving by William Hogarth, 1751 (Wellcome Library, London)

escape from their predicament and to maintain employment. In the second place, it trained young girls as domestic servants and boys to join the armed forces. In this way, subscribers to the Hospital funds hoped to reclaim and make useful otherwise neglected and unproductive members of society.

The rise of the hospital in so many guises (voluntary and state, general and specialist) created new spaces for the practice of medicine and new opportunities for doctors and nurses to advance their careers, develop new treatments and preach a gospel of hope for patients and their families. This is not to say that the medical profession became united in this period or that there were no disputes between practitioners. Professional disagreements persisted, particularly around issues such as status, fees and authority. Physicians questioned the rights of apothecaries to provide medical advice as well as dispense medication. Apothecaries in turn resisted the encroachment of chemists and druggists on their professional territory. Poaching patients from competitors was regarded in many countries as a form of malpractice and doctors' reputations were often challenged by patients and their peers, who were happy to expose unscrupulous, dangerous and deceitful practices. A surgeon could be forced to resign because he had failed to obtain a patient's consent to surgery.

Attempts to regulate disputes between doctors or between doctors and their patients included legislation designed to ensure adequate training and monitor entry into the profession, as well as the publication of advice books that set out the qualities of a good doctor (including abstinence from alcohol, according to some writers) and provided a framework for avoiding or resolving professional conflicts. Arguably the earliest British text on medical ethics was written by the Scottish physician John Gregory (1724–73), whose *Observations on the Duties and Offices of a Physician*, published in 1770, was based on the 'common sense' philosophy of Thomas Reid (1710–96). But perhaps the most well-known contribution to the field was made by Thomas Percival (1740–1804). Prompted by an internal dispute between colleagues in Manchester, in 1794 Percival published a guide to professional etiquette that was expanded in 1803 to become arguably the most influential Enlightenment discussion of professional medical ethics.

Regular or licensed practitioners both in the West and in China were often united, however, against a common enemy: unlicensed 'quacks'. The emergence of an increasingly global consumer society, with its emphasis not only on consumption but also on ownership and exchange, together with the limited

JAMES GRAHAM'S 'CELESTIAL BED'

Some quacks claimed to be able to solve sexual problems. The most renowned eighteenth-century 'sex therapist' was James Graham (1745–94), who lectured and treated patients in his 'Temple of Health' in London and promoted his 'Celestial Bed', available for hire at £50 per night, as a means of restoring health and vitality, reversing impotence and sterility, and ensuring 'ecstasy' as well as conception. The 'Celestial Bed' could earn Graham more in one night than most labourers earned in a year.

capacity of regular practitioners to cure or alleviate many diseases, created a market for alternative remedies. Quacks publicized their wares in newspapers and in pamphlets or handbills distributed on the street. Advertisements included words and images designed to attract customers to purchase pills and ointments that promised miraculous cures and the assurance of health. British proprietary medicines advertised in this way included Morison's Universal Pills, Bateman's Pectoral Drops for chest complaints and Godfrey's Cordial or Mother's Friend, which contained opium and was used to sedate (but sometimes killed) young infants.

Neither quacks nor the public accepted the denouncing of alternative approaches to health care. Orthodox practitioners were ridiculed by cartoonists as often as their unlicensed counterparts. Surgeons in particular were portrayed as torturers and 'butchers' or as the purveyors of pain and death. Acrimonious disputes between licensed and unlicensed practitioners, and indeed between the various classes of orthodox physicians, surgeons and apothecaries, were not immaterial. During the eighteenth century, all types of healers depended on attracting clients for their financial success or survival. Public reputation was critical and the stakes were high. Many licensed practitioners barely scraped a living but a successful and hard-working physician or surgeon, one who perhaps attended royalty and nobility or who, like William Hunter, displayed considerable entrepreneurial talent, might earn in excess of £10,000 per year. Such riches placed some doctors in the upper echelons of Enlightenment society.

Madness and midwifery

The Enlightenment veneration of reason led perhaps inevitably to the stigmatization and marginalization of people who exhibited various types of character or behaviour that were deemed to be irrational. Across Europe in particular, intolerance of insanity and superstition was evident in two aspects of contemporary life: an increasing reliance on institutions to care for the mentally ill, and shifting attitudes to the authority of women in relation to childbirth and childcare. While those suffering from madness and folly were increasingly regarded as dangerous and deviant because they transgressed the boundaries of acceptable behaviour set by civilized society, women were equated, at least in the minds of male commentators, with wilful and sometimes malign ignorance. Although attitudes to women and the mad were not uniformly dismissive, they reveal the impact of Enlightenment ideology on health policy as well as the ways in which medicine shaped social relations.

It is often assumed that before the seventeenth and eighteenth centuries the insane were cared for at home and 'village idiots' were allowed to wander freely in communities, at least as long as they were not violent. Understood largely in terms of divine or demonic possession or as the result of humoral imbalance, madness was generally not feared: the mad and foolish were not marginalized but accepted and sometimes celebrated. This caricature of pre-Enlightenment approaches to madness is not entirely misplaced. Institutions for the mad were certainly rare. Perhaps the oldest and most well-known hospital catering specifically for the insane was Bethlem (commonly referred to as Bedlam) in London. Founded as a religious institution for the sick in 1247, Bethlem admitted lunatics by the fourteenth century. Across Europe, similar places of refuge, known as 'asylums', were opened in Gheel, in Belgium, and in Valencia, Barcelona and other urban centres in Spain. As in England, these institutions were established and managed on a religious and charitable basis and built on the model of a caring community. Notwithstanding these overtly humane approaches, in late medieval, Renaissance and early modern Europe the mad were often regarded as a source of family shame and were segregated or hidden away from the rest of society.

The philosophical underpinnings of an Enlightenment shift to a more formal and repressive system of asylum care, referred to by the French philosopher and historian Michel Foucault (1926–84) as the 'great confinement', lay in the distinction between mind and body formulated by Descartes and in the theories of reason and knowledge developed by the English physician and philosopher John Locke (1632–1704). Unable to think clearly or rationally, or to work productively, the insane were seen as prominent symbols of the threat to social order generated by the spread of vagrants, paupers, criminals, beggars and prostitutes through the cities and countryside of modern civilized societies. Segregation in asylums and workhouses, particularly in France, Germany and England, was thus partly a mechanism for controlling social deviance and enabling local and state authorities in western Europe to incarcerate supposedly unproductive and dangerous members of the population. At the same time, the availability of specialist institutions promoted the emergence of 'alienists' or psychiatrists as a discrete sub-discipline within medicine and helped to generate new categories of mental disease.

As in previous periods, the mad were identified by their conduct. The stereotypical manic patient exhibited grandiose delusions, hallucinations, religious enthusiasm, bodily twitches, obsessive and repetitive actions and excessive sexual behaviour. The melancholic would be apathetic, distracted, idle and uninterested in friends and family. In either case, madness could result in suicide. Although insanity continued to be explained partly in functional humoral terms, more mechanistic models increasingly dominated medical accounts. Thus, madness was thought to be caused by the effects of injuries, fevers, illness and childbirth on the brain, as well as by emotional disturbances such as unrequited love. The dominance of bodily explanations linked to pregnancy and childbirth, and a growing belief in the inherent psychological instability of women, may help to explain why more women were diagnosed with insanity during the seventeenth and eighteenth centuries. This pattern was reinforced by historical and fictional examples of the horror and romance of female madness, evident in the revelatory chronicles of Margery Kempe (1373–c. 1438), whose mystical visions were sometimes dismissed as the product of insanity, and in the tragic literary figure of Ophelia.

Enlightenment doctors initially treated madness in much the same manner as many other diseases, focusing on restoring a balanced lifestyle, healthy diet and appropriate rest. With the rise of segregation in asylums, however, more brutal and coercive practices emerged. In some places, inmates were constrained by iron rings placed around their neck, body and arms in order to control violent tendencies and their physical and mental health was often neglected. Patients were treated with cold baths, electric shocks and isolation. In state institutions, and particularly in workhouses, patients were made to work, to reduce the financial burden of caring for them, and at Bethlem visitors paid to observe the spectacle of inmates chained like animals in pens. Private asylums exploited the insane in different ways. Driven largely by commercial interests, private madhouses offered families a means of removing unwanted and embarrassing, but often sane, relatives in order to gain control of their estates. In addition, private asylums provided little therapeutic support, serving merely as repositories for forgotten members of society.

Towards the end of the eighteenth century, these practices were challenged by the rise of 'moral therapy'. In 1774, concerns about abuse led the British government to insist on the licensing of all private madhouses and from 1800 such institutions had to have medical supervision. More widely, physical restraint came under scrutiny, as Enlightenment optimism began to inform care of the insane. In France, Philippe Pinel (1745–1826) introduced a variety of reforms at the Salpêtrière and Bicêtre hospitals, casting off the inmates' chains and emphasizing the role of more constructive and supportive psychological treatments. In England, a similar approach was initiated first at St Luke's Asylum by the physician William Battie (1703–76) and subsequently at the York Retreat, a charitable institution founded in 1796 by the Quaker merchant and philanthropist William Tuke (1732–1822). At York, Tuke and his associates pioneered an approach to madness that was shaped by idealized notions of wholesome domesticity, where recovery was promoted through the kindness, reason and humanity of doctors and the staff. Similar institutional reforms, which reflected political allegiance to liberty and equality as well as increasing scientific emphasis on the mind, rather than merely the brain, as the seat of mental disorders, were initiated elsewhere throughout Europe and North America.

There were other aspects of mental illness and its management in the eighteenth century. Although florid mania and melancholy were in many ways censured, the nervous instability described by George Cheyne in 1733 remained a fashionable disease that marked the affluent and sensitive middle classes apart from their less refined lower-class neighbours. The middle classes' obsessions with mental health combined with their increased spending power may also have led to the popularity of phrenology and mesmerism. Introduced by the German physician Franz Anton Mesmer (1734–1815), mesmerism involved the use of magnetic forces or hypnotism, especially to cure women with hysteria. Phrenologists such as Franz Joseph Gall (1758–1828) and JC Spurzheim (1776–1832) claimed that character, personality and mental ability could be revealed by close examination of the contours of the skull. Although these techniques were popular, the orthodox medical profession regarded mesmerism and phrenology as forms of quackery: both Mesmer and Gall were forced to leave Vienna and practise elsewhere.

Cheyne's preoccupations with overindulgence as one of the causes of nervous disorders resurfaced towards the end of the eighteenth century as a particular anxiety about the effects of alcohol abuse. Concerned partly by the impact of inebriety on the health of soldiers and sailors and by the availability of cheap liquor, Benjamin Rush and the Scottish-trained ship's surgeon Thomas Trotter (1760–1832) insisted that drunkenness was a mental disease that required medical treatment to prevent the catalogue of physical and psychological conditions caused by persistent drinking. Over-consumption of alcohol was thought to cause epilepsy, hysteria, inflammation of the eyes, liver and skin, apoplexy, gout, jaundice, indigestion, emaciation, diabetes, ulcers, impotence, premature ageing and madness. According to Trotter, the treatment of 'this vile habit' included the immediate withdrawal of alcohol and the prescription of specific remedies for physical symptoms.

Mental illness did not only afflict the poor. The madness of the British monarch George III (1738–1820) is well known. The precise cause of the king's periodic descent into delirium remains a subject for debate. According to some historians, George III was suffering from porphyria, a rare genetic disorder associated with hallucinations, anxiety and depression as well as a range of physical symptoms. For others, the madness of George III may have been a manifestation of a recurrent mood disorder, such as manic depression. Whatever the cause, the king's illness proved difficult to treat, either by the royal physicians or by a clergyman doctor, Francis Willis (1718–1807), who ran a private madhouse and whose services were requested when orthodox treatments proved ineffective. The failure of the British Prime Minister

William Pitt (1759–1806) to introduce Catholic emancipation in Ireland (which would have allowed Catholics to sit in Parliament) has been blamed on George III's irritability and irrationality during an episode of ill-health.

The demonization of irrationality and madness that marked Western approaches to mental health for much of the eighteenth century was arguably less evident in Oriental systems of health care. Chinese doctors had recognized mental illness (*k'uang*) for many centuries and Chinese medical texts described the wild behaviour of patients with 'mania' in very similar terms to Western accounts of madness. In Chinese medicine, however, understandings of mental illness were characterized by more holistic interpretations of the relationship between the mind and the body in shaping emotional health and by more nuanced accounts of an 'ethereal soul' that entered the body shortly after birth and flowed back to 'heaven' at the moment of death. Mental health, like physical well-being, continued to be understood in terms of the flow of energy and the balance between yin and yang. Less harsh and less invasive than Western asylum practice, Chinese treatments incorporated many of the traditional practices of yoga, acupuncture and herbal preparations to restore equilibrium.

Enlightenment intolerance of what was regarded as irrational was evident not only in the treatment of the insane but also in attitudes to women in many Western countries. Traditionally, a woman's place was in the home. In addition to contributing to the domestic economy and managing the household, women were responsible for contraception and the regulation of fertility, for the provision of care and support during labour and childbirth, and for nursing children and sick relatives. In some situations, women were also employed as wet-nurses to suckle the children of women who could not or did not want to breastfeed. Female friends and relatives continued to support their neighbours in an informal manner. In addition, families, Poor Law institutions and charitable hospitals employed female nurses and midwives. These practitioners were often regulated by the Church; midwives, for example, were required to hold a licence from a bishop in order to practise. During the age of reason, however, female nurses and midwives were increasingly regarded, particularly by men, as immoral, profligate, superstitious and unscrupulous. Women were thought to lack the education, knowledge and skill to practise safely and effectively and in some cases they were branded witches.

Attempts to oust women from their traditional areas of expertise were part of an attempt by male doctors to promote their own professional and economic interests. Although some British midwives, such as Sarah Stone, were well educated in anatomy and had published their own textbooks of midwifery, the majority of midwives were excluded from the education necessary to acquire knowledge of scientific developments. Such midwives were mocked by surgeons and apothecaries who looked to boost their fees by expanding their practice to include the treatment of women during and after childbirth. Only men, doctors argued, possessed sufficient understanding of nature to ensure the survival of both the mother and the child. Doctors' claims to superiority were bolstered by the fact that women were not allowed to use instruments such as the obstetric forceps, which had probably been invented by the French surgeon and royal *accoucheur* Pierre Chamberlen (1560–1631), or one of his relatives, during the sixteenth or seventeenth century to ease difficult deliveries.

Male midwives did not escape either censure or ridicule. Satirical cartoons often mocked the involvement of men in a traditionally female role (see Figure 13). Nevertheless, the strategy adopted by male doctors proved relatively successful, as men began to displace women as experts in childbirth and childcare. The full force of the male medical attack on ignorant midwives and nurses is evident in the writings of William Cadogan, who was a physician at the Bristol Infirmary and honorary medical attendant at the Foundling Hospital in London. In *An Essay*

Figure 13 A caricature of a man-midwife, 1793 (Wellcome Library, London)

upon Nursing, first published in 1748 in the form of a letter to the governors of the Foundling Hospital, Cadogan argued that uneducated midwives and wet-nurses were damaging the health of young children by over-feeding them with rich and luxurious foods not fit for infant stomachs and by smothering them in too many clothes. Emphasizing his experiences as a father as well as his clinical expertise, Cadogan insisted that reason and knowledge would provide a safer and more natural approach to childcare than superstition and tradition. In many ways, Cadogan's tract exemplifies Enlightenment optimism in the capacity of science and medicine to improve the health and wealth of modern populations.

A medical Enlightenment?

It is difficult to determine whether substantial shifts in medical ideology during the Enlightenment had any impact on patterns of health and disease or on the experiences of most patients around the world. Declining mortality, increasing rates of fertility, and population growth may have owed more to increased wages, better nutrition and wider social and economic developments than to changes in medical theories and practice. Even the notion that there was a 'medical Enlightenment' is problematic. Change was gradual and piecemeal, as new ideas and treatments were gradually merged with older traditional approaches to health and well-being. In addition, it is important to remember that there was no single system of health care and no unified theory of disease. During the eighteenth century, medicine was marked by the wide variety of practitioners, theories and treatments that were available to patients and their families. Patients could choose a particular doctor or remedy according to individual preference and financial means.

Notwithstanding these difficulties of interpretation, it is possible to identify certain characteristic features of Enlightenment medicine. Medicine was an increasingly secular enterprise, shaped by the opportunities provided by a consumer society. Successful doctors were often those who demonstrated entrepreneurial spirit or an ability to exploit the fears and aspirations of communities living under the recurrent threat of disease and death. Professional concerns about financial survival encouraged orthodox practitioners to expand the range of services offered to their patients and to disparage the practices of quacks and charlatans. The Enlightenment veneration of reason marginalized women from practice, even to the point of preventing them from fulfilling roles traditionally reserved for women, such as the care of mothers and children. In a similar vein, the mad were increasingly isolated in purpose-built state or privately-owned madhouses.

In spite of concerns that civilization was causing the proliferation of nervous diseases or the spread of infections, doctors and their patients were optimistic about the capacity of medicine to improve their lives. Enlightenment hopes were bolstered by developments in surgery, the introduction of vaccination, the growth of charitable welfare institutions and the emergence of specialist sites for medical education and practice. Together these factors provided the foundations for new understandings of anatomy and pathology, based not on humoral theory or on the presence of symptoms alone but on scientific knowledge about the structure and function of organs. After the tumultuous events of the French Revolution, these changes gathered momentum as the principal site for generating medical knowledge shifted away from the patient's bedside and became increasingly located first in the hospital and subsequently in the laboratory.

5

Science and surgery: medicine in the nineteenth century

> But since the antiseptic treatment has been brought into full operation, and wounds and abscesses no longer poison the atmosphere with putrid exhalations, my wards, though in other respects under precisely the same circumstances as before, have completely changed their character; so that during the last nine months not a single instance of pyaemia, hospital gangrene, or erysipelas has occurred in them.
>
> Joseph Lister, *British Medical Journal*, 1867

Saving lives can often be simple. Before the mid-nineteenth century, many women died from puerperal fever, a form of infection that was contracted during childbirth. The precise cause of puerperal fever remained unknown. According to some doctors, it was the result of germs spread from one person to another, in this instance from the doctor to the patient. For other physicians, fever and death were caused by contamination from the over-crowded and insanitary environment of the hospital. In Britain and America, doctors such as Alexander Gordon (1752–99) and Oliver Wendell Holmes (1809–94) tried to prevent the spread of infection by encouraging colleagues to change their clothes regularly and to wash with chlorinated water but their efforts had little immediate impact on the lives of young mothers and their families.

The problem of puerperal fever was explored in a more systematic way by the Hungarian-born physician Ignaz Philipp Semmelweis (1818–65), who worked in the maternity clinic at the *Allgemeines Krankenhaus* (General Hospital) in Vienna during the 1840s. Semmelweis noticed that mortality rates in the ward where babies were delivered by medical students and doctors were nearly ten times higher than in the ward run by midwives; approximately 29% in the former compared with 3% in the latter. Suspecting that doctors and their students were transferring 'putrid particles' from the dissection room to the obstetric ward, in 1847 Semmelweis ordered the doctors (and the midwives) to wash their hands in chlorinated lime before delivering babies. The mortality rates fell immediately. Although Semmelweis's simple instructions saved the lives of many Viennese women, his views were not universally accepted by his colleagues and he left Vienna for his home town of Budapest, where he successfully introduced similar antiseptic procedures in the maternity wards at St Rochus Hospital, leading some to regard him as the 'saviour of mothers'. In 1865, however, Semmelweis became mentally ill and was admitted to an asylum, where he died from a condition that had been induced either by over-work or perhaps by exposure to the same form of infection that he had helped to control.

Semmelweis's experiences reveal several ambiguities in the relationship between social and medical progress, on the one hand, and the health and well-being of European populations on the other. The modern industrial age clearly offered fresh opportunities, particularly to people seeking work and pleasure in expanding towns and cities. Innovative approaches to education, higher levels of employment, faster travel and communication, scientific progress and political emancipation combined to generate greater freedom and improve social and economic prospects. In addition, as the history of puerperal fever demonstrates, relatively straightforward innovations carried the potential to improve the likelihood of surviving disease and preventing despair. At the same time, however, dramatic social and technological changes brought new threats to health and welfare. Crowded urban slums, rising levels of atmospheric pollution, new forms of work and warfare and prolonged periods of economic depression facilitated the spread of infectious diseases as well as rising levels of mental illness, particularly amongst the vulnerable lower social classes. Living in a city was associated with reduced life expectancy, a phenomenon known as the 'urban penalty', as death rates from tuberculosis, cholera, dysentery and smallpox remained high. Poverty also compromised the health of isolated rural populations, who had limited access to health care or social support.

There were other paradoxes associated with progress. For many, the solution to seemingly intractable social problems lay in improvements in science and medicine. Refinements in surgery, new forms of medical education and the emergence of state medical support generated optimism in the power of progress to improve lives. These developments were not confined to Europe. Transmitted by travelling physicians or Christian missionaries, the theories and techniques of western European medicine (including surgery and anaesthesia) were increasingly adopted by Islamic, Chinese, Japanese and American doctors and their patients. Yet, even these medical developments introduced new hazards. Although medicine was usually hailed as a means of curing or preventing disease, it was also sometimes a cause of illness and death. Physicians, surgeons and nurses spread infections, patients died during surgery, and powerful drugs such as laudanum or arsenic could have fatal side-effects. Often parodied as unnecessarily cavalier in their approach to treatment, doctors working in hospitals and workhouses or for the armed forces were sometimes feared more than they were revered.

These diverse, and often conflicting, features of medicine and science in the nineteenth century are clearly evident in many contemporary contexts: in growing concerns about the role of dirt in causing disease and death; in disputes about the effects of anaesthesia and antisepsis on the practice of surgery and midwifery; and in the rise of laboratory science as a means of identifying more accurately the organisms or germs responsible for infectious diseases. But questions remain. What impact did developments in medical knowledge and practice have on the health and behaviour of people around the world? How did the social, cultural and environmental changes that were associated with modern industrialization influence patterns of disease and the practice of medicine? To what extent did alternative approaches to health care continue to attract interest from patients and their families? How effective were the growth of public health systems and the rise of laboratory sciences in reducing mortality from infectious diseases?

Pathology and public health

There is more than one way to die. As the forces of industrialization swept through western Europe, workplaces became more hazardous. Workers were injured or killed in industrial accidents, or suffered and died from diseases caused by repeated exposure to dirt and dust at work. Lung conditions such as silicosis and asbestosis, which led to shortness of breath, fatigue, fever, heart failure and sometimes cancer, were common amongst miners. Young chimney sweeps developed scrotal carcinoma, a form of cancer first described by the English surgeon Percivall Pott (1714–88) in 1775, caused by chronic irritation of the skin by soot particles. During the middle decades of the nineteenth century, governments in many countries introduced legislation that was intended to reduce occupational deaths by enforcing stricter regulation of working environments and establishing safer limits on working hours. In Britain, the Factory Acts of 1833 and 1844 reduced the hours that children were allowed to work and set up a system of routine factory inspection.

The rise of occupational diseases was not the only risk of modernization. Medicine also had the power to kill. For many centuries, opium had been routinely prescribed by doctors and consumed by patients as an effective means of combating pain and inducing sleep. Imported into western Europe, predominantly from Turkey but also from China and India, opium became increasingly popular as both a therapeutic and recreational drug during the nineteenth century. According to Mrs Beeton's *Book of Household Management*, published in 1861, parents should keep opium powder and laudanum (a tincture of opium) in their medicine cupboards to treat minor illnesses. While opium clearly offered hope and pleasure to many, treatment with opium, or with one of its derivatives such as morphine, could sometimes be worse than the disease for which it was prescribed. As the confessions of the English 'opium-eater' Thomas De Quincey (1785–1859) made clear in 1821, both patients and doctors could become addicted to opiates. Infants given laudanum, which was often included in soothing cordials designed to suppress a cough or to help the baby sleep, were also killed by overdoses. By the 1840s, disputes about the trade in opium between China and the British Empire (known as the Opium Wars) had contributed to widespread anxieties about the spread of 'opium dens', in which healthy men and women were thought to be in danger of being corrupted by Chinese vices. Objections to opium abuse led to the founding of the 'Anti-Opium Society', as well as attempts by politicians and the medical profession to regulate the sale of opium. In Britain, the Pharmacy Act of 1868 legally required opium to be labelled as a poison and introduced fines for failing to comply with the regulations. In spite of these measures, preparations of opium continued to be widely available from druggists and chemists.

Other medicinal drugs posed threats to health. Arsenic was a common ingredient of patent medicines sold for the treatment of cancer or skin conditions and was regularly used in everyday life, either as a means of killing vermin or as a dye in paint and wallpaper. It was also, however, a convenient method of poisoning wealthy relatives for personal gain, since it was extremely difficult to detect in a body after death. Plants containing strychnine had been used by medical practitioners in China and India for many centuries and were employed in Europe in the nineteenth century as pesticides or in the treatment of traumatic shock, for example following a railway accident. Like arsenic, strychnine was also fatal in overdose, whether administered deliberately or in error: strychnine poisoning caused vomiting, convulsions and asphyxia. Fears of the unregulated sale of hazardous medicinal substances and their occasional use as murder weapons encouraged greater legislative control over their availability and prescription. After 1868, British chemists were required to keep a record of sales of arsenic, cyanide, mercury and strychnine, including details of the buyer and the purpose for which these substances were purchased.

THE TRIAL OF WILLIAM PALMER

In 1856, the British surgeon William Palmer was indicted for the murders of his gambling associate John Parsons Cook, his wife Anne Palmer and his brother Walter Palmer for financial gain. Although only small amounts of antimony were discovered in the bodies, the celebrated pathologist Alfred Swaine Taylor (1806–80) testified at the inquest and trial that Palmer had poisoned his victims with strychnine, using his medical knowledge to ensure that the poison was not detectable. The trial of William Palmer attracted national attention, demonstrating contemporary fears about the abuse of dangerous drugs and the growing importance of medical testimony in court. Palmer was found guilty of Cook's murder and hanged.

If drugs comprised one aspect of the dangers of medicine, the rise of pathological anatomy constituted another potential threat to life. In the clinics and dissection rooms of hospitals, particularly in Paris, anatomists attempted to identify associations between the symptoms of disease detected during life and the findings from post-mortem examination. These approaches to human disease were aided not only by greater technical skill in dissection but also by innovations in clinical diagnosis. In 1819, the French clinician and pathologist René Théophile Hyacinthe Laennec (1781–1826), who had been trained in the arts of diagnosis and pathological anatomy by Corvisart, published an account of how his newly invented stethoscope helped to correlate breath sounds with post-mortem findings, leading to new classifications of cardiac and respiratory disease. Two years later, Laennec's work was translated into English by the Scottish physician Sir John Forbes (1787–1861), leading to the widespread adoption of the stethoscope across Europe and North America.

The revolutionary French style of pathological inquiry was disseminated rapidly, as students from around the world travelled to Paris to acquire new diagnostic and anatomical skills and wrote their own dissection manuals. There was, however, a dark side to this late Enlightenment project. Greater reliance on anatomical knowledge led to an ever-increasing demand for bodies for dissection. Traditionally, in many European countries, surgeons and anatomists were entitled to dissect the bodies of executed criminals but this source was no longer sufficient to satisfy the requirements of new schools of anatomy and medicine. As a result, doctors began to buy bodies that had been disinterred from recent graves, leading to a lucrative trade in human flesh. In Britain, gangs of body-snatchers and grave-robbers could charge up to twenty guineas for a fresh body, a figure far in excess of the average urban wage of around thirty shillings a week in the early nineteenth century.

The most infamous manifestation of this macabre practice was the wilful murder of poor vagrants. In the late 1820s, two Irish immigrants from Ulster, William Burke and William Hare, killed sixteen people and sold their bodies to an Edinburgh surgeon, Robert Knox (1791–1862), for approximately £10 each. When the body of the final victim, a woman from Donegal, was discovered in a lodging house and the two men were charged with murder, Hare gave evidence against Burke and escaped punishment. Burke was sentenced to death and hanged in January 1829. Although he avoided official censure, Knox's role in the affair was widely criticized by the public and his peers and it severely damaged his career. One of the immediate consequences of the trial of the 'trading assassins', as they became known, was the passage of the Anatomy Act of 1832, which allowed surgeons to dissect the cadavers of people who had died in hospitals or workhouses and whose bodies remained unclaimed by relatives. The Anatomy Act provided a supply of corpses for the London surgeon and anatomist Henry Gray (1827–61), the author of *Gray's Anatomy*, which was first published in 1858 and became the most influential anatomy textbook since the *Fabrica*. The Act also transferred the indignity of dissection from convicted criminals to the poor.

Although death by drugs and dissection generated sensational media coverage, it would be wrong to overemphasize the dangers that medicine posed to nineteenth-century people. There were greater threats to health and happiness. Chronic conditions, such as heart disease, cancer, rheumatism and diabetes, continued to afflict Western populations as they had done in previous centuries. Although speculations about the causes and mechanisms of heart disease remained tentative, during the nineteenth century cardiac conditions such as angina pectoris, aortic aneurysm, atheroma, arteriosclerosis (hardening of the arteries), fatty changes in the heart and myocardial infarction were increasingly carefully described in terms of their clinical presentation and autopsy findings. There was also considerable interest in cancer, with clearer scientific classifications of leukaemia and lymphoma, for example, and more detailed descriptions of the cellular processes by which

HEREDITARY HEART DISEASE

Heart disease was a major cause of sudden death known to run in families. On 11th June 1842, Thomas Arnold (1795–1842), the headmaster of Rugby School, suffered a severe attack of angina and died. The post-mortem revealed structural changes to the heart but no disease of the valves or pericardium. A similar condition had killed Arnold's father and in 1888 was responsible for the death of one of his sons, the poet Matthew Arnold (1822–88).

cancerous growths invaded adjacent tissues and spread around the body. The only available treatment for most forms of cancer was surgery but survival rates were extremely low, often in the region of only five percent.

Nineteenth-century developments in clinical examination and post-mortem investigations resulted in original descriptions of many modern conditions. In 1817, the London physician James Parkinson (1755–1824) published his observations on the 'shaking palsy', now known as Parkinson's disease. Illustrating his account with patients from his own practice, Parkinson described the gradual appearance of the tremors, and the muscle weakness, posture and gait that came to be regarded as characteristic features of the illness. Some years later, Thomas Addison (1793–1860), a physician at Guy's Hospital in London, described the signs and symptoms of pernicious anaemia, a condition caused by vitamin B12 deficiency, as well as the fever, fatigue, weight loss and weakness caused by damage to the adrenal glands, a condition that became known as Addison's disease. Addison was not the only Guy's physician in this period to have a disease named after him. Bright's disease (of the kidneys) was named after one of the most famous and most influential doctors in England, Richard Bright (1789–1858). And one of Addison and Bright's colleagues at Guy's, Thomas Hodgkin (1798–1866), described the enlargement of the spleen and lymph glands that characterized the condition subsequently called Hodgkin's disease or Hodgkin's lymphoma. It is no coincidence that Addison, Bright and Hodgkin had all graduated from Edinburgh, which had become one of the leading centres of Enlightenment medicine.

In some cases, chronic diseases continued to be regarded as the result of immorality or immoderate lifestyles, leading to a social stigma evident particularly in the cruel, but not uncommon, condemnation of obesity. In other cases, chronic illness was understood as the product of emotional stress caused by adverse life circumstances such as bereavement, ill-health, marital discord or financial problems. From the mid-nineteenth century, increased mortality from heart disease was often explained in terms of the stress and strain of modern professional life and the hurry and bustle created by urban over-crowding and new forms of high-speed travel and communication, such as the railway and the telegraph. Diabetes and thyroid diseases were similarly regarded as the product of prolonged business pressures and emotional disturbances. Increasing scientific interest in chronic disease generated new knowledge. In 1848, a London physician, Alfred Baring Garrod (1819–1907), demonstrated the presence of uric acid in the blood and joints of patients with gout, leading eventually to new methods of diagnosing gout and approaches to treatment that combined specific dietary restrictions with the use of old-fashioned anti-gout drugs such as colchicine, derived from the autumn crocus or meadow saffron.

There are some intriguing contemporary personal accounts of the impact of chronic ill health on patients and their families. In *Life in the Sick-Room*, published in 1844, and in a variety of essays, the English writer Harriet Martineau (1802–76) revealed not only the manner in which pain, fatigue, deafness, anxiety, indigestion and possibly heart disease had confined her to a life of chronic suffering and invalidity but also her belief in the curative power of mesmerism and other unorthodox therapies. The post-mortem examination, an account of which was published in the *British Medical Journal* in 1877, confirmed that Martineau had a large ovarian tumour or cyst in her abdomen but concluded that her death had been caused by 'failure of the heart's action'. Other writers, such as Jane Austen (1775–1817), Charles Dickens (1812–70), the Russian physician and dramatist Anton Chekhov (1860–1904), the Spanish novelist Benito Pérez Galdós (1843–1920), the French short story writer Guy de Maupassant (1850–93), and the American author Charlotte Perkins Gilman (1860–1935), included accounts of illness and the behaviour of the medical profession in their novels, plays and short stories.

Epidemics and pandemics of infectious disease continued to ravage populations around the world. Indeed, in the case of certain diseases, these became even more potent and more feared causes of death than they had been in previous centuries. The transmission of tuberculosis, typhoid, typhus, smallpox and influenza was facilitated by the polluted slums and sewers of over-crowded cities, sometimes referred to as 'urban graveyards'. Certain groups were more vulnerable than others. Official statistics reveal the extent to which the young and old succumbed more readily to all forms of infectious disease and the manner in which typhus, a condition particularly associated with poverty and dirt, became a major cause of death amongst prisoners and soldiers. Combined with cold and hunger, the spread of typhus through the French army in 1812 was largely responsible for the humiliating failure of Napoleon's invasion of Russia.

The disease that perhaps best captures the perils of urban living in the nineteenth century is cholera, a condition that, like typhus, was associated with high military mortality. In the Crimean War (1854–6), the American Civil War (1861–5) and the Boer War (1899–1902), far more people died of cholera, typhus and dysentery than were killed by enemy weapons. Cholera was probably first described as a distinct condition, separate from other forms of fever, in India during the late eighteenth century. From India, the disease travelled along trade routes to China, the Philippines, Russia and Turkey. By 1831, cholera had reached western Europe, from where it was subsequently transported across the Atlantic by ship, first to Canada and North America, before eventually spreading further south to Mexico and beyond.

It is not surprising that cholera was feared. Patients suffered excruciating pain, vomiting, diarrhoea, dehydration and rapid death. The speed and impact of the spread of cholera across the world was revealed by new approaches to recording the incidence of disease and death and by the application of statistical methods to the resulting data. In England, approximately 22,000 people died during the first six months of the 1831–2 epidemic, 7,000 of them in London. Recorded deaths from cholera in Britain between 1831 and 1866 were in the region of 140,000. During the epidemics between 1847 and 1861 over one million Russians died from cholera. High mortality rates led to over-crowded graveyards and further fears of the uncontrollable spread of disease from bodies buried hurriedly too near the surface. It was partly in response to widespread concerns about the environmental pollution caused by infected dead bodies that cremation began to become a popular alternative to burial.

Reactions to cholera varied. In Britain, the United States and Germany, cholera epidemics prompted the creation of municipal Boards of Health set up to oversee efforts to curtail the spread of the disease. The Cholera Prevention Act, passed in Britain in 1832, empowered local authorities to provide medical and nursing support to cholera-stricken communities, to destroy infected clothes and bedding and to cover drains and sewers. In the United States, sanitarian reformers lobbied for higher taxes to improve public health services and reduce water pollution. There were, however, disagreements about the best methods of prevention and management. Those who believed that the source of infection lay in polluted water and air (the 'miasmatists') emphasized the importance of sanitary reform, insisting that slums should be cleared and filth removed from streets and rivers. For those who regarded infection as spread predominantly through human contact (the 'contagionists'), the isolation or quarantine of infected patients or travellers was the preferred method of control. In some places, both approaches were employed in valiant attempts to halt the devastation caused by cholera and other infectious diseases. However, as satirical images by Robert Cruikshank (1789–1856) made clear, individuals and communities remained perplexed by medical advice and unsure about the most effective means of preventing or curing the disease (see Figure 14).

Figure 14 Caricature of a cholera patient experimenting with remedies, Robert Cruikshank, 1832 (Wellcome Library, London)

In Britain, attempts to improve urban sanitation following the first cholera epidemic owed much to the energy and insights of Edwin Chadwick (1800–90), a lawyer and author of the 1842 *Report on the Sanitary Condition of the Labouring Population of Great Britain*. Chadwick was instrumental in the introduction of the New Poor Law in 1834. Based on the belief that poverty was a product of idleness rather than socio-economic conditions, the New Poor Law established workhouses as a means of ensuring that the destitute worked for their living and provided a deterrent to the idle, undeserving poor. As secretary to the newly created Poor Law Commissioners, Chadwick became increasingly aware of the relationship between poverty and disease, leading him to investigate the sanitary condition of Britain and to map patterns of disease over time and across space. Chadwick believed that infectious diseases were the product of environmental filth and dirt. His proposals for major engineering initiatives designed to provide clean water to households and to enable the removal of sewage provided the basis for modern public health.

Chadwick's report was influential beyond Britain, stimulating similar investigations and sanitary reforms in North America and Germany. During the 1830s and 1840s, the strategies employed by national authorities to reduce the spread of infectious diseases across Europe and America continued to vary according to local conditions and political regimes. However, these initial public health measures were challenged in 1848 by a second cholera pandemic, which proved more damaging than the first. In response to fears of an imminent epidemic, the British government introduced the Public Health Act of 1848. The Act established a Central Board of Health, as well as encouraging local authorities to appoint Medical Officers of Health responsible for improving sanitary conditions. These provisions proved ineffective as cholera swept through the population once again, killing more than 50,000 people. Similar cholera epidemics decimated communities in North America in 1849–55, 1863–6 and 1890; in some cities death rates were over three percent.

Disputes about the precise mode of transmission of cholera persisted until the results of investigations carried out by the London physician John Snow (1813–58) became clear. Snow charted the precise pattern of cholera cases following an outbreak in London in 1853–4 and demonstrated that a spate of deaths in the Broad Street area could be traced to afflicted families' use of a specific local well. Further outbreaks of cholera in the district were prevented by removing the handle from the Broad Street pump, an intervention that served to restrict exposure to drinking water contaminated with faeces. Snow's investigations effectively proved the role of infected water in transmitting cholera and partially vindicated Chadwick's environmentalist approaches to health and disease. Environmental factors were also stressed by the Bavarian chemist Max Joseph von Pettenkofer (1818–1901), who in 1892 famously drank 'cholera water', to disprove the presence of germs, and led to the adoption of comparable sanitary measures in Germany. However, subsequent studies by Snow suggested that the disease was caused by a living organism, which was eventually identified by the German doctor and bacteriologist Robert Koch (1843–1910) in 1884.

Cholera was not the only epidemic disease to threaten global populations during the middle decades of the nineteenth century. Communities around the world were also vulnerable to outbreaks of smallpox, influenza, diphtheria, yellow fever and many other infectious diseases. In China, epidemics of malaria, cholera, dysentery and trachoma (a disease of the eyes caused by chlamydia infection) were regular occurrences and encouraged the adoption of Western scientific methods of disease control. Although vaccination against smallpox was made compulsory in Germany in 1807 and in Britain in 1853, opposition to the practice left many children and adults unprotected. As a result, epidemics swept through Europe, particularly at times of conflict. During the Franco-Prussian War of 1870–71, smallpox killed both German soldiers and French prisoners of war. The disease was probably carried to Britain by French refugees, and killed more than 44,000 people. One of the consequences of epidemics of smallpox was the establishment of specialist hospitals, such as the charitable London Smallpox and Vaccination Hospital.

Major epidemics of influenza also occurred throughout the nineteenth century, as they had in previous centuries, particularly during the winter months. In general, fatality rates were low but during the first influenza pandemic, between 1889 and 1892, known as the 'Russian flu' because of its origins in St Petersburg, over 300,000 people in Europe died from the disease. Meteorological and medical records from those years suggest that increased morbidity and mortality during outbreaks may well have been triggered by specific environmental conditions. Hospital admissions and sickness absence amongst workers at the General Post Office in London both increased three or four days after severe and sudden drops in temperature. Epidemics of influenza and other infectious diseases led to the establishment of special fever or isolation hospitals, such as those built in Glasgow, Dublin and Manchester.

Public health officials and military leaders throughout Europe were also anxious to stop the spread of venereal diseases, such as syphilis and gonorrhoea, which were thought to be undermining the strength and vitality of armies and navies. From the 1860s, Britain, France, Germany, Sweden and other European countries attempted either to ban or regulate prostitution more successfully or introduced compulsory medical inspection of prostitutes in ports and garrison towns, forcing them to be treated in specialist hospitals if they were found to show signs of disease. These measures were unpopular in many places because of the manner in which they infringed civil liberties and targeted women, rather than men. By the 1880s, most legislation had been repealed and alternative methods of controlling sexually transmitted diseases were pursued. Although sometimes dismissed on religious and moral grounds, attempts to prevent the spread of venereal diseases (and to limit pregnancies) were enhanced by the appearance of more effective condoms following the introduction of vulcanized rubber in 1844.

Municipal and private initiatives to combat infectious diseases were combined with advice to families about the importance of personal cleanliness. During cholera epidemics, John Snow advised people not only to avoid drinking contaminated water but also to keep patients isolated and to ensure that families and visitors washed regularly. The working classes, in particular, were instructed to keep their bodies and homes clean and respectable. From the 1840s, public baths and wash-houses were built in some areas of Britain, to promote cleanliness amongst the inhabitants of urban slums. In middle-class homes, the bedroom washstand was gradually replaced by wooden, tin and cast-iron baths, situated in carefully designed bathrooms. By the end of the nineteenth century, public health education included advice about the principles of household hygiene, the preservation of food, and the removal of dust and dirt from the home by the use of disinfectants and vacuum cleaners. Family members were advised to wash their hands before meals or to shave off beards to reduce the risk of infection. Like moderation in the consumption of food and drink, cleanliness was advocated not only for its health benefits but also for religious and moral reasons. Clean bodies and homes were thought to indicate moral purity.

HOOVER

The first electric vacuum cleaner was manufactured by William H. Hoover (1849–1932). In 1907, James Murray Spangler (1848–1915), a department store janitor in Ohio, had invented a 'carpet cleaner and sweeper' in an attempt to control his dust-induced asthma. Hoover bought the patent, established the Electric Suction Sweeper Company and began production of his Model 0 vacuum cleaner in 1908.

More systematic approaches to public health and hygiene were not confined to the West. European policies for preventing the spread of infections were transported to South Asia, particularly to the provinces of British India, where sanitary measures were implemented primarily to protect the health of armies and British settlers. Western models of modernization in medicine and science also became popular amongst some Chinese officials during the 1860s. Chinese students increasingly travelled abroad to study medicine or were able to receive formal training in the theories and practices of Western medicine at state institutions such as the Tianjin Medical School, which opened in 1881. However, westernization was not routinely accepted. Many medical practitioners and state officials in India and China remained resistant to the marginalization of indigenous forms of health care and the gradual adoption of Western forms of medical education. In Japan, acceptance of Western medicine varied regionally. In Tokyo, students were increasingly educated in scientific medicine imported from Germany. In Kyoto, by contrast, physicians such as Akashi Hiroakira (1839–1910) taught and administered Western medical treatments alongside traditional Japanese remedies.

The rise of state-sponsored public health in Europe, North America and the Far East had an impact on the professional boundaries of medicine. To ensure the quality and value of health care, governments began to insist that doctors working as Medical Officers of Health or as Poor Law doctors should be suitably qualified and licensed to practise. New standards of education and ethics were expected. Government pressure to create a professional body of doctors who would carry out civic duties in the face of epidemic diseases coincided with concerns within the orthodox or regular profession about competition from unorthodox or alternative practitioners. In Britain, reform of the profession involved setting up a system of registration and monitoring of practice that excluded those designated as quacks: the homeopaths, herbalists, hydropaths, osteopaths, chiropractors, phrenologists, spiritual healers and others who continued to provide health care and advice to people still struggling with the ever-present threat of death and despair. The British Medical Act of 1858 introduced a register of all licensed doctors and established the General Medical Council, responsible for overseeing education and standards of practice. In the United States, by contrast, the medical marketplace remained more fluid. Regular practitioners established their authority over their rivals by emphasizing their clinical superiority. The success with which orthodox practitioners united their profession against competition from irregulars was bolstered by developments in pain relief and infection control that provided the basis for renewed optimism in the power of medicine to cure disease and alleviate discomfort.

Anaesthesia and antisepsis

In 1828, a London surgeon, Bransby Blake Cooper (1792–1853), removed a stone from the bladder of a 53-year-old labourer from Sussex. Instead of the customary six minutes, the operation lasted nearly an hour and the patient died the following day. Many of Cooper's peers were unimpressed and his surgical incompetence was exposed by the editor of *The Lancet*, Thomas Wakley (1795–1862), who suggested that Cooper had only been appointed to his position at Guy's Hospital because he was a nephew of one of the most prominent and distinguished surgeons and anatomists of the period, Sir Astley Cooper (1768–1841), who had taught the poet John Keats (1795–1821). Bransby Cooper sued Wakley for libel. Although Cooper won the case, the jury largely endorsed Wakley's allegations, by awarding Cooper only £100 in damages rather than the £2000 that he had claimed.

The case of Bransby Cooper illustrates several features of early nineteenth-century medicine. First, it highlights the potential for patronage, or in this case nepotism, to secure lucrative hospital appointments for friends and relatives. Second, it demonstrates the power that doctors held over the fortunes of poor patients and their families. Offered free treatment in charitable hospitals, the poor were in no position either to argue with or reject medical opinion or practice. Of course, not even wealth precluded exposure to clinical arrogance and brusqueness: the English surgeon John Abernethy (1764–1831) was quite happy to insult his private patients, advising overweight men to eat less and their overweight wives to exercise more. But the poor were particularly vulnerable to medical conceit. Finally, the patient's death at Cooper's hands testifies to the dangers of surgery in this period. Even relatively routine operations, such as lithotomy or amputation, involved considerable risk of pain, infection, haemorrhage and death.

During the middle decades of the nineteenth century, surgery was gradually transformed by the introduction of more effective anaesthesia and the application of antiseptic and aseptic techniques in operating theatres. In earlier centuries, surgery had been made bearable by the use of alcohol, opium and mandrake to deaden the pain, and by the speed of the surgeon: from the first incision to the final stitch and dressing, it took the London surgeon Robert Liston (1794–1847) only three and a half minutes to amputate a leg. From the early nineteenth century, combinations of drugs were sometimes administered to induce sleep and facilitate surgery. The Japanese surgeon Hanaoka Seishū (1760–1835) produced a mixture of herbal substances, which he referred to as *tsūsensan*, that contained many of the drugs used today in general anaesthesia, such as scopolamine, hyoscine and atropine. Seishū demonstrated the efficacy and safety of *tsūsensan* by using it to perform amputations and remove breast tumours.

During the 1840s, new anaesthetics became available. General anaesthetics, such as nitrous oxide (laughing gas) and ether, were first used in dentistry. In 1842, an American medical student, William E. Clarke (1819–98), anaesthetized a patient with ether in order to extract a tooth. Two years later, Horace Wells (1815–48) demonstrated the efficacy of nitrous oxide first by having one of his own teeth removed and subsequently by giving public displays of tooth extraction under anaesthetic. Perhaps the most significant experiment with an anaesthetic was carried out in October 1846 by William T.G. Morton (1819–68), a dentist and medical student in Massachusetts. Having used it in his dental practice, Morton administered ether to allow a surgical colleague to remove a tumour from the neck of a young man. News of Morton's success spread, leading to the rapid adoption of ether as an anaesthetic throughout North America and Europe.

Widespread recognition of the benefits of ether was followed only a year later by the introduction of another anaesthetic agent, chloroform. The ability of chloroform to induce sleep was discovered accidentally by the Scottish surgeon and obstetrician James Young Simpson (1811–70) in 1847, when he was experimenting with various chemical substances with two friends. Simpson began to use chloroform anaesthesia to reduce women's pain during childbirth. Interest in chloroform intensified after 1853, when John Snow safely administered it to Queen Victoria during the birth of Prince Leopold. New and safer methods of delivering chloroform and other inhaled anaesthetics were developed by Joseph T. Clover (1825–82), an English doctor who administered anaesthetics to several thousand patients, including prominent public figures such as Florence Nightingale (1820–1910) and the Prime Minister Sir Robert Peel (1788–1850), without any deaths (see Figure 15).

Acceptance of ether and chloroform by doctors and their patients was encouraged not only by the willingness of members of the monarchy to undergo anaesthesia but also by evidence of the value of pain relief on the battlefield. During the Crimean War, many amputations were successfully performed under anaesthesia. In addition, effective anaesthesia allowed orthodox practitioners to marginalize alternative approaches to controlling

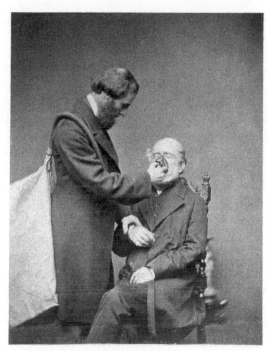

Figure 15   Joseph T. Clover demonstrating his chloroform inhaler on a patient, 1862 (Wellcome Library, London)

pain, such as mesmerism and hypnotism. However, it would be a mistake to assume that anaesthesia was adopted without dispute. Within medicine, a number of objections were raised against the use of ether and chloroform. According to some religious writers, pain during childbirth was necessary and natural and should not be alleviated by artificial means. Like other forms of suffering, physical anguish was regarded as a valuable part of human experience, until the rise of humanitarian commitments to alleviating pain served to undermine belief in the spiritual meaning of pain.

For some critics, inhaled anaesthetics were dangerous. In addition to the risk of explosions, during the late 1840s and

1850s there were reports of patients experiencing severe nausea or suffering heart attacks under anaesthesia. Doctors were also anxious about the impact of experimental surgery on their reputation. Anaesthesia made it possible to perform more invasive surgery on unconscious lower-class patients, and particularly on women, leading to potential abuses of power. Wider public fears of anaesthesia related to the use of ether and chloroform by criminals to subdue victims of theft, rape and murder. In spite of these reservations, however, the accumulation of evidence that anaesthesia provided safer operations led to its gradual acceptance as a routine part of hospital care, allowing surgeons to become less brutal and more skilful. A reduced reliance on speed and strength may eventually have encouraged more women to become surgeons. By providing a safer and less frenzied operating environment, anaesthesia also hastened acceptance of the second major nineteenth-century innovation in medical care, antisepsis.

According to certain historians, 1865 was a pivotal year in the emergence of modern medical science. This was the moment when the English surgeon Joseph Lister (1827–1912) introduced antiseptic techniques into surgical practice. Before the mid-nineteenth century, approximately half of patients could expect to die during or after an operation because of infection and blood loss. During the 1840s, Semmelweis's efforts to reduce deaths from puerperal fever were reinforced by Thomas Spencer Wells (1818–97), who became surgeon to Queen Victoria in 1863 and was subsequently president of the Royal College of Surgeons of England. In addition to advocating washing in cold water, Wells introduced the practice of using fresh towels during surgery and prohibited students and doctors from attending an operation if they had performed an autopsy within the previous week. Lister was therefore by no means the first person to attempt to eradicate the risk of infection during surgery.

Lister's antiseptic method was based on a belief that surgical infections were caused by exposure to 'minute organisms' present in dirty hospital environments. Building on previous attempts to reduce the frequency of infections by applying alcohol or carbolic acid to wounds, Lister began to use a carbolic acid spray in theatre to disinfect the surgeon's hands, the instruments, the wound and the air around the patient. He recorded his first success with the technique in August 1865, when he treated a young boy's fractured leg by dressing it in bandages soaked in carbolic acid. Normally, such an injury would become infected and require amputation. In this case, the wound healed and the boy recovered without losing his leg. In 1867, Lister published the results of further operations using his antiseptic method of preventing infection. His figures demonstrated the remarkable improvements in survival rates since the introduction of antisepsis: the death rate following amputation, for example, fell from 45% to 15%.

As in the case of anaesthesia, antisepsis was not immediately accepted by Lister's colleagues, who complained that his methods were overly complicated and unnecessary and that improved surgical outcomes could be attributed to better hospital ventilation, healthier diets and more attentive nursing. Nevertheless, experiences of managing wound infections during the Crimean and Franco-Prussian wars supported Lister's approach and encouraged gradual acceptance of the principles of antisepsis. By the late 1870s, Lister's techniques were employed by many surgeons across Europe. In subsequent years, Lister's emphasis on killing germs present at the site of a wound or operation was replaced by efforts to prevent contamination in the first place, an approach referred to as asepsis. Aseptic techniques included sterilizing equipment and clothing with heat, disinfecting the hands of theatre staff and eventually the adoption of surgical masks and rubber gloves.

By the end of the nineteenth century, the widespread availability of anaesthesia and asepsis had transformed surgery from an experimental procedure likely to end in death into one that patients could realistically expect to survive. More invasive forms of surgery became possible. Ovaries and wombs were removed, teeth extracted, tumours excised and parts of the bowel cut out with greater confidence. The operative skill of the Scottish surgeon Sir William Fergusson (1808–77), who had been a pupil of the infamous anatomist Robert Knox, made it possible to repair cleft lips and palates. Occasionally, surgical optimism spilled over into exploitation and abuse of professional power. American surgeons practised on slaves, and both British and American women were subjected to mutilating surgery as a means of enforcing control over behaviour that was considered abnormal or unseemly. Normal ovaries were taken out in an attempt to treat hysteria or neurosis and in the 1860s an English surgeon, Isaac Baker Brown (1812–73), performed clitoridectomies to cure nymphomania or prevent female masturbation. Brown was heavily criticized by his peers for operating without consent and was expelled from the Obstetrical Society of London.

Women were disadvantaged in other ways by social values and professional developments. Although Florence Nightingale and the Jamaican Mary Jane Seacole (1805–81) became famous for reducing mortality rates and providing nursing care for injured soldiers in military hospitals during the Crimean War, women remained largely excluded from medical practice. Only rarely were women such as Elizabeth Blackwell (1821–1910), Elizabeth Garrett Anderson (1836–1917) and Mary Putman Jacobi (1842–1906) admitted to European or American medical schools and allowed to practise medicine, teach medical students and conduct research. Obstacles to the entry of women to the profession included concerns that they were not strong enough, the belief that education and work would reduce their ability to have children, and fears that the medical marketplace would become even more crowded. According to an editorial in *The Lancet* in 1873, women were 'neither physically nor morally qualified for many of the onerous, important, and confidential duties of the general practitioner'. Such judgements reflected wider cultural anxieties about female emancipation and were vigorously rejected by early pioneers of the feminist movement such as Sophia Jex-Blake (1840–1912), who qualified in medicine in Switzerland and subsequently founded medical schools for women in London and Edinburgh.

In spite of attempts to reinforce the male monopoly on medical care, women began to enter the health-care professions, first as nurses and midwives but increasingly as doctors. Both nurses and midwives had frequently been denigrated as ignorant and disreputable but in the nineteenth century reformers in many countries strove to convert nursing into a respectable profession, particularly for young middle-class women. In London, Florence Nightingale introduced reforms in recruitment, education and practice that produced a hierarchical system of management, within which working-class women were largely employed as nurses on the wards while middle-class women became ward sisters and matrons. Nurses were expected to conform to an ideal standard in terms of education, character and appearance, although the reality in many places may have been different. Midwives too attempted to improve their professional status. In Japan, improvements in midwifery services were part of a more general set of social reforms during the Meiji period (1868–1912). Japanese midwives were forbidden from performing abortions or selling medication and were increasingly expected to undergo training, examination and registration in order to practise.

In certain countries, such as Sweden, Switzerland and France, women were gradually entitled to study medicine from the 1860s and a small number of British women travelled abroad to obtain qualifications. Nevertheless, women faced opposition, and sometimes physical violence, from male staff and students, and the roles and responsibilities of women within medicine were clearly circumscribed. Women doctors were thought to be particularly suited to the care and treatment of children and other women; that is, to aspects of family and community life that were regarded as traditionally female. Women's autonomy to practise in other health-care domains was also challenged. During the late nineteenth century, massage was often prescribed by doctors but administered by women as a means of restoring strength to weak and injured limbs or of providing rehabilitative exercises to those fatigued by nervous exhaustion. In Britain, suspicions that prostitutes were masquerading as masseuses and concerns about the propriety of women touching men's bodies led to attempts to bring massage under stricter medical regulation and to distance remedial care more explicitly from sex work. The supposed dangers of massage were sensationalized by the orthodox medical profession in the pages of the *British Medical Journal* in 1894, leading to attempts by remedial therapists to professionalize their practice with the formation of the Society of Trained Masseuses, the forerunner of the Chartered Society of Physiotherapy.

Greater regulation of the boundaries of acceptable practice did not prevent the continuing popularity of alternative approaches to health care, particularly amongst families who were either unable to pay for conventional treatments or unwilling to expose themselves to the risks of scientific medicine. Various proprietary medicines, such as Beecham's Pills for constipation, stomach ache and 'female disorders', became popular in Britain. Although often ridiculed by orthodox physicians, homeopathy also proved fashionable. Homeopathic medicine was devised by the German doctor Samuel Hahnemann (1755–1843) in the late eighteenth century and was based on two principles. First, Hahnemann argued that substances known to cause specific symptoms in healthy people could be employed to relieve those same symptoms in patients, a principle that he referred to as 'like cures like', which echoed the ancient Eastern concept of correspondence. Second, Hahnemann believed that cure was effected by infinitesimal doses of the medicine, leading him to prescribe exceedingly dilute preparations of substances in order to restore balance and health to the ailing body. Taught in dedicated schools across Europe, America and India, homeopathy appealed particularly to middle- and upper-class patients who could afford to avoid the unpleasant side-effects and potentially perilous consequences of orthodox medicine.

Chinese techniques were also adopted by European doctors and their patients. Although British doctors initially criticized Chinese medicine for its apparent lack of anatomical knowledge, their French counterparts were more receptive. Gradually, however, even sceptical British physicians began to accept the practice, if not the theory or cosmological implications, of acupuncture and employed it as a relatively safe method for treating a wide range of conditions, such as chronic pain, convulsions, gout, rheumatism, dysentery and eye diseases. In the West, heterodox or complementary approaches to health were increasingly challenged in the late nineteenth century, as licensed doctors effectively pushed alternative practitioners to the fringes of medicine. Exclusion was supposedly justified by referring to the growing authority of modern medical science. In the laboratories of universities and medical schools in Germany, Britain, France and America, scientists were beginning to develop new understandings of health and disease and to generate novel approaches to prevention and treatment that would transform medical practice.

Fighting germs

The modern scientific laboratory was arguably a German invention. During the early and mid-nineteenth century, universities at Breslau, Heidelberg, Leipzig and Munich established physiological laboratories, which were sometimes located in specific institutes, designed to generate new knowledge that would contribute to improved medical care, and to provide spaces and facilities for educating students. As this model of education and research spread to other institutions, American and European medical students travelled in large numbers to German universities to acquaint themselves with new experimental techniques and acquire new skills. In particular, students learned the art of vivisection, in which physiological knowledge was acquired from the dissection of live animals. In addition, they pursued investigations into the chemical and histological properties of bodily organs in health and disease, thereby providing more detailed descriptions of metabolic processes and innovative diagnostic tests.

After the Franco-Prussian War and the unification of Germany in 1871, the pursuit of scientific knowledge for practical purposes (referred to as *realwissenschaft*) became a key component of German attempts to achieve technological, military, scientific and economic supremacy. Government investment in higher education became a pivotal mechanism in the pursuit of national pre-eminence. The industrial and commercial benefits of laboratory science rapidly became apparent. During the second half of the nineteenth century, German chemical dye companies had begun to diversify into the investigation and purification of new medicines. In the 1890s, one German company, Friedrich Bayer and Company, synthesized and marketed a derivative of salicylic acid, under the trade name Aspirin. Derived from willow bark, the medicinal properties of which had been known since antiquity, salicylic acid had traditionally been used to reduce fever and pain. The new derivative was less irritating and rapidly became popular, particularly during epidemics of influenza, generating considerable profits for Bayer. After the First World War, however, Bayer's revenues declined as the company's assets and rights were seized by the American government as part of post-war reparations. Eventually, the patent for Aspirin expired and cheaper generic forms of the drug became available.

The development of German scientific laboratories owed much not only to private and state investment in research facilities but also to the pioneering zeal of leading scientists. The experiments of chemists such as Justus von Liebig (1803–73) in Giessen, physiologists such as Johannes Müller (1801–58) in Berlin and Carl Friedrich Wilhelm Ludwig (1816–95) in Leipzig revealed many features of bodily structure and function, particularly in relation to the actions of the neuromuscular, circulatory and respiratory systems. Observations and experiments were made possible by improved microscopes, such as those made by the German mechanic Carl Zeiss (1816–88) in Jena, and by novel equipment, including the kymograph that recorded arterial pressure or muscle contractions. Many of Müller's students continued his system of physiological investigation. As well as identifying the role of an enzyme, pepsin, in digestion, Theodor Schwann (1810–82) used microscopy to develop the theory that all animals and plants were composed of cells. The pathologist Friedrich Gustav Jacob Henle (1809–85) published detailed accounts of the histological structure of the eye, brain and kidney. The loop of Henle in the kidney, which ensures effective concentration of the urine, is one of the anatomical structures named after him.

Perhaps the most famous of Müller's students was the physician and pathologist Rudolf Virchow (1821–1902), who became professor of pathological anatomy in Würzburg before returning to Berlin in 1856 as director of the Pathological Institute. Virchow rejected contemporary beliefs in spontaneous generation, a theory that postulated that organisms could develop from inanimate matter, by demonstrating that all cells originated in other cells, a concept encapsulated in the phrase 'omnis cellula e cellula'. Cancer, inflammation and reproduction, he argued, could all be explained in terms of the normal or abnormal behaviour of cells. In addition, Virchow shaped the development of pathology, training the next generation of scientists and founding and editing a leading international journal on pathology, physiology and clinical medicine. Virchow was not only an accomplished and influential pathologist but also keenly committed to social and political reform as a means of improving public health.

The German model of scientific investigation was continued in the work of Robert Koch (1843–1910). Influenced by Müller, Henle and Virchow, Koch studied medicine in Göttingen, worked as a general practitioner for ten years and then became a scientist in the Imperial Department of Health in Berlin. Koch's contributions to the emergent field of bacteriology were extensive and demonstrate the close relationship in this period between human and veterinary medicine. He identified the organism responsible for anthrax, a potentially fatal disease of animals that could be transmitted to farm workers. In addition, Koch isolated the organisms that caused tuberculosis (1882) and cholera (1884) and published studies of the causes of wound infections. In the early 1890s, Koch announced the development of a vaccine (a term now used to describe any form of inoculation) against tuberculosis, one of the most prominent causes of death across the world at that time. Unfortunately, Koch's belief in the value of tuberculin as a means of preventing the disease proved unfounded and created considerable scandal, partly because of occasional deaths following vaccination and partly because Koch attempted to benefit financially from his innovation. Nevertheless, Koch's research provided crucial supporting evidence for the theory that specific diseases were caused by specific organisms, thereby contributing to what became known as the germ theory of disease and to the development of specific cures.

German laboratory science clearly provided a model for the establishment of institutes and laboratories around the world, including the Pasteur Institute in Paris (1887), the Lister Institute in London (1891) and various research institutions in North America funded by philanthropists such as John D. Rockefeller (1839–1937). However, parallel alternative traditions of scientific investigation emerged in other countries. French experimental physiology owed much to the work of François Magendie (1783–1855) and Claude Bernard (1813–78). Famous for his work on the vasomotor nerves, the mechanisms of neuromuscular transmission, the role of pancreatic secretions and the function of the liver in glucose metabolism, Bernard formulated what became a pivotal concept within physiology, namely the theory that the stability of the internal bodily environment was essential for physiological health: '*La fixité du milieu intérieur*', he wrote in a passage that became one of the most celebrated in modern physiology, '*est la condition de la vie libre, indépendante*'. Bernard's studies provided the framework for the notion of 'homeostasis' (or physiological equilibrium) subsequently developed by the American physician and physiologist Walter B. Cannon (1871–1945).

Bernard's notoriety was perhaps eclipsed by one of his French compatriots, the chemist Louis Pasteur (1822–95), who spent most of his scientific career in Paris. Pasteur's contributions to modern science and medicine are almost too numerous to mention. His early studies revealed the role of yeast in fermentation and confirmed Virchow's rejection of spontaneous generation. In subsequent work, Pasteur demonstrated that heating wine, beer and milk (a process now referred to as 'pasteurization') would eliminate the organisms responsible for souring those products. Pasteur adapted his experimental procedures to study the importance of micro-organisms in putrefaction and in the generation of diseases in both animals and humans. His most celebrated successes included his role in identifying the organism that was infecting silkworms and threatening to destroy the French silk industry, and his development of vaccines for the treatment of chicken cholera, anthrax and rabies. Although rabies was not common, painful death from the disease was inevitable. Experimenting on rabbits, Pasteur developed an inactive form of the virus, which he tested on a nine-year-old boy who had been bitten by a rabid dog. The vaccine successfully halted the development of the disease and Pasteur's treatment became widely accepted around the world as the safest and most effective means

PASTEUR

Figure 16   Lithograph of Louis Pasteur, celebrating his work on rabies, 1880s (Wellcome Library, London)

of treating rabies (see Figure 16). It was largely as a result of his research on rabies that the Pasteur Institute, funded by both state and philanthropic donations, was established in 1887.

The success of Pasteur's theories and therapies owed much both to his ability to popularize his work through public demonstrations and to his oratorical and entrepreneurial skills. The acceptance of germ theories of disease was also shaped by the remarkable success of the treatments that followed. In the last decade of the nineteenth century, the recognition that serum from infected patients and immunized animals carried protective properties raised hopes that infectious diseases might be treated by the administration of specific antitoxins or serum containing antibodies. The first demonstration of the clinical value of serum therapy, or passive vaccination, was achieved by Emil von Behring (1854–1917) and Shibasaburō Kitasatō (1852–1931). Working in Robert Koch's laboratory at the Institute of Hygiene in Berlin, von Behring and Kitasatō raised diphtheria antitoxin by injecting animals with sub-lethal doses of purified toxin. On Christmas Day in 1891, they successfully used the antitoxin to treat a child suffering from diphtheria. This novel technique for combating the ravages of infectious diseases, for which von Behring received the first Nobel Prize in medicine in 1901, was rapidly applied not only to larger groups of children with diphtheria but also to patients with tetanus. The impact of serum therapy was immediately clear. The commercial production of diphtheria antitoxin and its introduction to clinical practice in various European hospitals initiated a dramatic and sustained reduction in mortality rates from diphtheria.

Germ theories generated other innovations, particularly in relation to the treatment of specific diseases, by producing what the German scientist Paul Ehrlich (1854–1915) referred to as 'magic bullets' that would target specific germs. Ehrlich's work led to the introduction and widespread prescription of 'salvarsan', which replaced mercury for the treatment of syphilis, especially during and after the First World War. In conjunction with the work of Pasteur, Koch and others, Ehrlich's innovations ushered in a golden age of bacteriology and immunology, an age characterized by significant optimism in the power of medical science to discover new treatments and transform the natural history of disease. The success of the bacteriological approach was also evident in the identification and management of tropical diseases such as malaria, yellow fever and sleeping sickness. The growth of tropical medicine as a speciality was promoted largely by the need to protect the health of missionaries, administrators and soldiers in the colonies. Designated schools of tropical medicine, such as those in London and Liverpool, were established in Europe and the United States. In many ways, it was medicine that made colonization and imperial expansion possible, not only for the British but also for other European and North American powers.

The success of laboratory medicine in preventing or treating epidemic infectious diseases was tempered by problems and disputes. Deaths from immunizations were not uncommon, leading many to regard the practice as experimental and dangerous. In some cases, fatalities were the result of severe allergic reactions to the animal serum that was injected. In others, disease may have been transmitted with the serum. Opposition to vivisection, which led in Britain to legislation regulating the use of animals in scientific research, also generated controversy about the ethics of experimental procedures. Nevertheless, clinicians increasingly embraced science as the basis for professional identity and authority. Education in science became a key requisite for qualification as an orthodox practitioner; clinical studies at medical school were increasingly preceded by two years of 'pre-clinical' education in the basic medical sciences. Chairs in physiology, anatomy and pathology were created in university medical schools in Europe and North America, and knowledge of the latest scientific innovations was disseminated through journals, learned societies and international congresses. One consequence of this growing reliance on science was a widening of the divide between orthodox and alternative medicine.

There was some resistance to scientific medicine, particularly from older-style clinicians, such as the Canadian physician William Osler (1849–1919), who regarded clinical medicine as more of an art than a science. In addition, there was a darker side to the optimism associated with new means of identifying individual diseases and with the emergence of new forms of knowledge. The belief that specific diseases had specific causes and that modern science could distinguish objectively and reliably between the normal and the pathological was applied to other areas of social life. Predicated on belief in the diagnostic and therapeutic power of biomedical science, new classifications of mental illness and biological formulations of racial difference were employed to support the marginalization or segregation of people regarded as 'unfit'. In Europe and America, large asylums were built to house expanding numbers of people diagnosed with insanity who were unable to contribute to industrial productivity or who transgressed behavioural norms.

Throughout Europe and North America, fears that an identifiable urban underclass of deviant degenerates was undermining the health and wealth of nations also led to policies that legitimized the compulsory sterilization or permanent institutional care of 'mental defectives', who were regarded as criminal, promiscuous and disruptive. Strategies for segregating those labelled defective were supported by statistical studies of the inheritance of criminality and intelligence carried out by the Italian criminologist Cesare Lombroso (1835–1909) and the British statisticians Francis Galton (1822–1911) and Karl Pearson (1857–1936). In addition, genealogical studies of families in which criminality, alcoholism and mental deficiency appeared to be more common reinforced beliefs in the inheritance of deviance. Attempts to draw a clear boundary between normal and pathological behaviour promoted the identification of homosexuals as deviant and legitimized the emergence of eugenics and racial hygiene that culminated in the Nazi extermination of Jewish people during the 1930s and 1940s. While many people undoubtedly benefited from myriad developments in medicine during the nineteenth century, the consequences of scientific and technological progress were not always as humane as contemporary reformers claimed.

Health trends and transitions

Between approximately 1870 and 1950, mortality rates declined and the life expectancy of many populations around the world increased. During the same period, while morbidity and mortality rates from infectious diseases fell, those from chronic diseases rose. There have been considerable disputes amongst historians about the scale and distribution of this demographic and epidemiological transition. While some historians have focused on declining trends in overall mortality rates, others have emphasized the manner in which infant and maternal death rates, as well as illnesses and deaths amongst the lower social classes, remained relatively high at least until the mid-twentieth century. In addition, as several historians have pointed out, mortality rates amongst non-Western populations remained significantly higher than amongst their Western counterparts throughout the twentieth century. It was not until the 1960s that life expectancy in India reached approximately 50, a figure achieved in Britain sixty years earlier.

Alongside arguments about differential trends in mortality and life expectancy, there have been disputes about the causes of the epidemiological transition. One of the earliest explanations, offered most frequently by medical practitioners, was that declining death rates amongst modern populations could be attributed to major medical innovations, including anaesthesia, antisepsis, immunization, the development of antibiotics, and the identification and treatment of deficiency diseases such as rickets and diabetes. In some cases, there may be evidence to support this interpretation. It is likely that maternal mortality rates fell dramatically after the mass production of antibiotics facilitated the treatment of puerperal fever caused by high rates of infection following illegal backstreet abortions. In addition, the isolation of insulin in 1922 contributed substantially to the management of diabetes, reducing both the risk of diabetic coma and the development of cardiovascular complications.

After the Second World War, however, a number of scholars began to question the story of medical progress embedded in popular accounts of declining mortality rates. In the 1950s and 1960s, Thomas McKeown (1912–88), professor of social medicine at the University of Birmingham, argued that preoccupations with medical advances were misconceived. He pointed out that lowered mortality from tuberculosis, perhaps the biggest killer in the late nineteenth century across the world, preceded the identification of germs and the development of specific preventative or therapeutic strategies. Although McKeown accepted that public health measures, including sanitary reform, contributed to the fall in infectious diseases and improved life expectancy, he suggested that the primary factors involved in shaping mortality trends were social rather than merely technical. Modern populations, according to McKeown, were rendered more resistant to infectious diseases by rising standards of living, better wages and improved nutrition.

McKeown's hypothesis has been challenged in recent years. Social historians of medicine have criticized his focus on tuberculosis at the expense of other diseases, disputed his tendency to define medicine rather narrowly in terms of scientific and technological advances, and pointed out that there is little direct evidence for his hypothesis that nutrition was the most important factor in reducing the impact of infectious diseases. Current scholarship suggests that a complex mix of inter-related sociocultural, political and scientific developments served to extend the life expectancy of modern populations. These mid-twentieth-century debates about the historical relationship between health and medicine were not merely about the past. They also reflected ongoing disputes about health inequalities and the availability and accessibility of health-care resources in the modern world. As the history of twentieth-century medicine demonstrates, while clinical practice may have become more dependent on science and technology, global health challenges and the strategies used to tackle them continued to be determined by social, political and economic factors.

6

War and welfare:
the modern world

> One does not need a crystal ball to predict that, within this
> generation, medical science will have overcome, and controlled,
> all man's external enemies.
>
> Ritchie Calder, *Medicine and Man*, 1958

In 1967, a brave new medical world was born. On the night of 2
December that year, Christiaan Barnard (1922–2001), a cardiac
surgeon at the Groote Schuur Hospital in Cape Town, performed
the first human heart transplant. Although the recipient, Louis
Washkansky, lived for only eighteen more days, Barnard's second
patient survived over eighteen months, demonstrating the feasi-
bility of transplanting the most important organ in the body.
During the 1970s and 1980s, the transplant of hearts became a
standard operation as surgical techniques were improved, heart-
lung machines became more effective, recipient and donor
matching was made more precise, and new drugs were invented
that could suppress immunological rejection of the organ. By the
end of the twentieth century, around the world, approximately
5,000 hearts a year were being transplanted and 70% of adult
recipients were surviving more than five years after transplant.
One patient survived for thirty-one years until his death in 2009.

Although iconic, Barnard's pioneering operation was not the
only medical success of the post-war period. During what has

been regarded as a golden age of medicine, between the 1950s and 1980s, kidney and lung transplants, hip replacements and endoscopic keyhole surgery became routine; more potent medicines were introduced to treat infections and mental illness; effective contraception became available; and a range of sophisticated tools, such as lasers, ultrasound, computerized tomography (CT) and magnetic resonance imaging (MRI), transformed the art of diagnosis and facilitated the early detection and treatment of disease. The capacity of doctors to assist couples who were unable to have children was also significantly enhanced. The birth, on 25 July 1978, of Louise Brown, the world's first test-tube baby, was a fitting climax to three decades of remarkable biomedical creativity.

Such developments were regularly sensationalized in the popular press, which detailed every scientific and clinical innovation and every miraculous story of creation and survival. Medical triumphs were not, however, without their critics. Inspired by the writings of the Austrian philosopher and social critic Ivan Illich (1926–2002), the Hungarian psychiatrist Thomas Szasz (1920–2012), and the Scottish psychiatrist R.D. Laing (1927–89), many disgruntled commentators denounced the growing authority with which medicine governed all aspects of everyday life. Modern scientific medicine, they argued, had overreached itself. New medical technologies were ethically contentious, physically and psychologically damaging and potentially fatal. According to Illich, medicine was often the cause, rather than the cure, of disease: an expanding proportion of symptoms, he argued, were 'doctor-made, or *iatrogenic*'. Illich's claims were supported by investigations that revealed the human and economic cost of unnecessary operations and seemingly avoidable deaths in the Western world.

Other aspects of Western biomedicine were challenged. Szasz and Laing claimed that the boundaries of mental illness were being expanded to justify greater social control, leaving modern populations addicted to antidepressants and anti-anxiety medicines. Increased investment in technological and pharmaceutical solutions to health-care problems was also thought to be diverting resources from attempts to reduce poverty or tackle the social causes of health and disease. Too much attention was being paid to curing specific diseases and too little to promoting mental, physical and social health. As many critics pointed out, the inhabitants of developing countries and those in some regions and racial groups in the developed world continued to experience deprivation, famine, insanitary housing, polluted water supplies, high levels of infectious diseases and inadequate access to modern public health care. The absence of wealth often determined the absence of health. From this perspective, the golden age of medicine was by no means equitable.

The complex and changing relationship between war and medicine, the evolution of welfare services across regions, nations and the world, and the rise of biotechnology during the twentieth century raise crucial questions about the risks and benefits of medical progress and about the impact of social change on patterns, understandings and experiences of disease. How have political, military and economic forces shaped modern health-care services, both locally and globally? What impact have science and technology had on the preservation of health and the treatment of disease? To what extent have developments in biomedical science altered the expectations and experiences of patients and their families? What are the limits of modern medicine?

War and medicine

In his reflections on the medical history of the Second World War, published in 1953, the American physiologist and historian of medicine John F. Fulton (1899–1960) suggested that 'the most significant advances in medical science' were made by military and civilian doctors working under the pressures of war. Fulton's claim is not without merit. Since antiquity, the battleground has been a theatre for the development and implementation of new surgical skills and techniques, the introduction of novel approaches to the management of burns and wounds and the treatment of infections, and the dissemination of improvements in sanitation, rehabilitation, education and nursing practice. War has also served as an incentive for the creation of global humanitarian welfare organizations, such as the Red Cross and the World Health Organization (WHO), and better systems for distributing health-care services and transporting casualties. These successes have been recognized by modern societies: during the twentieth century, many of the scientists and clinicians responsible for wartime innovations were awarded a Nobel Prize in Physiology or Medicine.

It is clear, however, that the relationship between war and medicine is more complex than this simple medical audit implies. The scientific and clinical outcomes of warfare have varied considerably according to the nature of the conflict and the specific branch of medicine concerned. The reputation of doctors has sometimes been tarnished by their commitment to military, rather than medical, goals and the health and safety of soldiers has been compromised by their being unknowingly subjected to experimental procedures. Advances made during wartime have not always been translated into civilian practice, and the requisitioning of national resources during war may have left certain civilian populations vulnerable to new threats to health. From these perspectives, war has not always been good for medicine or indeed for the health and welfare of patients.

The Crimean, American Civil and Franco-Prussian wars were instrumental in the promotion of nursing reforms and in the identification of functional heart disease, such as 'irritable heart', a term introduced in 1871 by Jacob da Costa (1833–1900) to describe the palpitations, increased pulse, chest pain and respiratory distress experienced by soldiers under duress. It was, however, during the Boer War and the First and Second World Wars that the power and limits of medicine became evident in the face of the devastation wrought not only by disease but also by more destructive forms of combat. During the Boer War (1899–1902) deaths from infectious diseases were double those due to injuries and wounds. British soldiers were especially vulnerable to typhoid, which killed thousands and against which there appeared to be little protection, contributing to the humiliating defeats suffered at the hands of the Boer forces. Nevertheless, new technologies and treatments were introduced during the conflicts in South Africa, including the application of field dressings, the appointment of dentists, the development of conservative abdominal surgery and the use of X-rays to locate bullets. In addition, wartime experiences against the Boers stimulated the subsequent introduction of inoculation against typhoid.

The Boer War had other consequences for the British people. One of the key problems exposed by the British press, highlighted by an official inquiry into the losses in South Africa, was the poor physical state of army recruits. According to some statistics, nearly sixty percent of men were rejected for service because of disability or ill-health. Fears about fitness were compounded by anxieties about the impact of urban slums on health and by concerns that the population was degenerating, leading to increasing levels of mental deficiency, tuberculosis, insanity, criminality and promiscuity. High infant and maternal mortality rates were thought to be threatening Britain's military and industrial authority on the world stage, particularly in comparison with the growing power of a unified Germany and the increasing commercial strength of France, Italy and the United States. Given the importance of the Empire to British supremacy and self-identity, the health of the nation became a matter of considerable political importance.

Initiatives aimed at redressing Britain's declining imperial and colonial authority were shaped by a belief, neatly expressed in the 1920s by the Canadian physician Helen MacMurchy (1862–1953): 'No baby, no nation.' Early twentieth-century child welfare programmes were partly modelled on the French *gouttes de lait* (milk depots), which had successfully reduced infant mortality from diarrhoea by encouraging breast-feeding and supplying sterilized milk for babies. Similar schemes were introduced in Britain and the United States as part of a strategy to reduce infant deaths and promote national efficiency. The success of milk depots was reinforced by the use of health visitors to educate mothers about child-rearing and by the establishment of more formal state mechanisms for monitoring and improving child health. In Britain, indirect outcomes of the Boer War included the introduction of free school meals in 1906 and the creation of the School Medical Service, which was responsible for inspecting and treating school children, in 1907. Although parallel attempts were made to improve the care of women during pregnancy and labour and to increase infant survival by requiring that midwives were suitably trained and that all births were notified, political momentum for the reform of maternity services was lost during the First World War.

The Great War was a watershed in the history of combat medicine. Between 1914 and 1918, approximately eight million soldiers were killed in action, twenty million returned wounded and seven million were listed as missing. Partly as a result of more destructive weapons and the perils of trench warfare, and partly because of improvements in the prevention and treatment of infections, for the first time in military history the number of casualties and deaths from physical injuries exceeded those from disease and starvation. In addition, the bodily wounds inflicted by cannons, mortars and aerial bombing were partially overshadowed, certainly in political significance if not in scale, by the proliferation of mental disturbances, or shell shock, amongst men in active service. Recent estimates from France, Germany and Britain suggest that in each country approximately 200,000 soldiers returned home suffering from some form of psychological disorder. In Britain, 58,402 men (6.8% of all disabled servicemen) received a war pension for shell shock and a further 44,855 for 'functional diseases of the heart'.

The First World War was the incentive for a number of innovations in medicine and surgery. Treatment was transformed by the development of routines to sort casualties according to the severity of their injuries, by new methods of cleaning wounds and irrigating them with disinfectants and by the provision of early medical attention in advance dressing stations (see Figure 17). X-ray machines were used to visualize the location of shrapnel, detect gas gangrene and identify fractures more accurately. Soldiers and civilians injured during bombing raids benefited

Figure 17 Advance dressing station in the field during the First World War (Wellcome Library, London)

from treatment in specialist fracture clinics and casualty departments, where new methods of diagnosis, management and rehabilitation could be provided. Improved vaccination programmes, mobile bacteriological units and delousing stations helped to reduce the incidence of typhoid, smallpox and typhus. Towards the end of the war, clinics aimed at tackling venereal diseases such as syphilis were established.

More strikingly, pioneering approaches to reconstructive surgery were developed by the New Zealand surgeon Sir Harold Gillies (1882–1960). Working first at the Cambridge Military Hospital in Aldershot and subsequently at Queen's Hospital in south-east London, Gillies focused not only on basic anatomical reconstruction but also on improving the aesthetic results of plastic surgery, using skin grafts and skin flaps to treat disfiguring facial wounds and burns. During the war, Gillies and his team operated on several thousand injured soldiers, paving the way for more ambitious plastic surgery in later years. The results of Gillies's surgical expertise were recorded in drawings made by the British surgeon and artist Henry Tonks (1862–1937).

The First World War also challenged the presumptions and practice of psychiatry. In the late nineteenth century, doctors across the world believed that it was middle- and upper-class women who were most vulnerable to nervous conditions such as hysteria and neurasthenia. This stereotype of mental illness as a 'female malady' was reinforced by clinical demonstrations of the manifestations and causes of hysteria by Jean-Martin Charcot (1825–93) and by theories of the sexual origins of neuroses promoted by the Austrian psychoanalyst Sigmund Freud (1856–1939). These assumptions were rudely shattered by the experiences of the First World War, when men of all classes and all nations broke down under the stress of combat. Although initially understood as the result of physical trauma caused by exploding shells, the physical and emotional symptoms of shell shock were subsequently attributed to the fatigue, sleeplessness, exhaustion, anxiety and fear generated by prolonged periods of conflict and hardship. The experiences of shell-shocked soldiers and the difficulties faced by families in adjusting to the return of their traumatized relatives after the war prompted doctors to acknowledge the presence of nervous conditions amongst men and the lower classes. Although awareness of shell shock stimulated the development of psychiatry and encouraged governments to screen future recruits more carefully for signs of mental vulnerability, many soldiers and their families continued to struggle with the burden of psychological damage generated by the First World War.

A number of administrative and organizational changes were implemented during the First World War. Motorized ambulances, aerial evacuation and radio communication facilitated early medical intervention. The creation of charitable convalescent homes, such as those opened after 1916 by the Royal Star and Garter charity in Britain, and the expansion of remedial massage or physiotherapy enhanced the likelihood of successful recovery and rehabilitation. In many cases, these developments in treatment and management continued to have an impact on health care after the war. Wartime losses served to re-energize efforts to address major public health issues and to improve infant and maternity welfare services throughout Europe and North America. In Britain, the Maternal and Child Welfare Act of 1918 placed responsibility for welfare provision on local authorities. The creation of the Ministry of Health in 1919 was intended to improve the co-ordination and delivery of health services across the country by supporting research, introducing public health measures, improving housing and nutrition and offering advice to hospitals.

The impact of the First World War on health and medicine, however, was not always constructive. The reputation of doctors was damaged by their role in identifying men as 'malingerers' and by frequent deaths from surgery. Although there is some evidence to suggest that civilian health may have improved for some people as the result of rising wages and better nutrition, civilian care probably declined in many areas and, apart from psychiatry and orthopaedics, wartime advances in specialized medical care were not necessarily adopted during peacetime. In addition, increasing numbers of men and women began to smoke cigarettes as a means of alleviating the tension associated with work and war, leading eventually to rising rates of chronic bronchitis and lung cancer. The sense of collective exhaustion at the end of the First World War may have encouraged the rapid spread of influenza, or Spanish flu, in 1918, a pandemic that killed approximately twenty million people worldwide. In the inter-war years, the cost of rehabilitating disabled soldiers and renewed demands for better services in the wake of the monumental loss of life placed considerable strain on health-care systems, particularly in the context of global economic recession and growing political unrest across Europe.

The close and reciprocal relationship between medical science, clinical practice and military success became apparent once again in 1939. During the Second World War, at least fifty million military personnel and civilians died as the result of combat, air raids, war atrocities, particularly those carried out by the Nazi and Japanese authorities, and the nuclear bombs dropped by US planes on Hiroshima and Nagasaki in 1945. As in the First World War, global warfare prompted the development and application of scientific, surgical and medical techniques, some of which were extensions of innovations introduced during earlier conflicts. Inoculation effectively prevented cases of tetanus, even in the squalid conditions of the evacuation of British forces from Dunkirk. Disease control measures, including spraying pesticides and providing soldiers with quinine, or its synthetic substitute mepacrine, served to minimize the impact of malaria on Allied troops fighting in Burma and North Africa. Blood transfusions and sulphonamide antibiotics improved the treatment of injuries and wounds and often reduced the need for life-threatening radical surgery.

Gillies's ground-breaking work in plastic surgery was continued by his cousin, Sir Archibald McIndoe (1900–60), who introduced innovative approaches to treating burns and encouraged the rehabilitation and reintegration of severely scarred patients.

Although 'flying stress' was recognized as a distinct condition, particularly amongst pilots in the British Bomber Command, armed forces personnel who broke down under the stress of war were often still stigmatized. Pilots who threatened operational efficiency by failing to cope with the fear and fatigue of battle were deemed to be deficient in character and, if found guilty of a 'lack of moral fibre', could either find themselves relegated to a basic grade or discharged. The pejorative tone of this label was not routinely endorsed, either by doctors or the armed forces. Nevertheless, as in previous approaches to shell shock, diagnoses of this nature were infused with moral judgement and led to the punishment and humiliation of many British service personnel. North American approaches to combat stress were more sympathetic. Emotional casualties were largely regarded by American and Canadian authorities as inevitable products of war, rather than as examples of lack of courage or malingering. It was thought that anyone could break down under sufficient stress and that social support and strong military leadership were more important than individual weakness in determining a soldier's capacity to cope with the stress of battle.

After the war, scientific and clinical interest in stress deepened. Military studies revealed the extent to which service personnel had struggled to cope with the pressures of combat, and the manner in which civilian morale and health had been damaged by bombing raids. Stress did not disappear on the return of peace. Long-term disability, the challenges of reintegrating war veterans into family life, and the problems faced by societies attempting to reconstruct lives and landscapes continued to render post-war populations vulnerable to psychological and physical breakdown. Subsequent military action promoted further debates about the causes and symptoms of combat stress. During the Korean War, scientists tried to establish the best means of predicting, measuring and preventing stress more effectively. After the Vietnam War, the prevalence of stress-related symptoms amongst American veterans led, in 1980, to the identification of post-traumatic stress disorder (PTSD), a condition that continued to plague military personnel and to generate considerable political debate during the closing decades of the twentieth century.

In spite of the range of medical innovations introduced between 1939 and 1945, accounts of medical advances during the Second World War have often focused heavily on the discovery of penicillin. The central features of the story are well known. In 1928, the Scottish biologist Alexander Fleming (1881–1955) noticed that the *Penicillium* mould that had accidentally contaminated his experimental cultures of bacteria at St Mary's Hospital in London had the capacity to kill microorganisms. Fleming named the antibacterial substance 'penicillin' but was unable to isolate the active component. Successful identification and production of penicillin was only achieved during the Second World War. Working in Oxford and funded by both the British and American governments, Howard Florey (1898–1968), Ernst Chain (1906–79) and Norman Heatley (1911–2004) produced sufficient penicillin to demonstrate its clinical efficacy. By the end of the war, mass production of penicillin in American laboratories enabled the treatment of thousands of infected Allied soldiers and heralded a new era of medicine, characterized by pharmaceutical commitment to developing other 'magic bullets' capable of targeting specific microorganisms. 'When I woke up just after dawn on September 28, 1928', Fleming commented many years later, 'I certainly didn't plan to revolutionize all medicine by discovering the world's first antibiotic, or bacteria killer but I suppose that was exactly what I did'.

Penicillin was not the only antibiotic to be produced during the war; streptomycin was introduced for the treatment of tuberculosis.

However, it is the production of penicillin that has usually been used by historians to support Fulton's claim that major scientific discoveries, or at least their application to clinical problems, have generally been inspired by wartime demand. Alongside other medical and surgical interventions, bacteriological discoveries certainly reduced casualties from wounds and infections during the Second World War. As in previous conflicts, however, doubts were expressed about the ethics of scientific investigation and military medical practice. Historical records show that soldiers were sometimes exploited as experimental subjects, often without their consent. The American authorities, for example, exposed soldiers to radiation and anthrax, to gain information about atomic and biological weapons. Such experiments were not without precedent. In the inter-war years, vulnerable groups such as children, the 'mentally defective' and prisoners had been used as human guinea pigs to test new drugs. In the infamous Tuskegee Syphilis Study, which began in 1932, the United States Public Health Service withheld treatment from black men in order to investigate the natural history of the condition.

After the Second World War, these experimental approaches to generating scientific knowledge were increasingly regarded as unethical. International outrage at experiments carried out not only by Nazi doctors such as Josef Mengele (1911–79) but also by other Axis and Allied forces led to attempts to regulate human experimentation. The Nuremberg Code of 1947 and the Declaration of Helsinki of 1964 established international guidelines for ethical scientific investigation and clinical practice, including the need to obtain subjects' voluntary consent. In addition, in 1948 the World Medical Association's Declaration of Geneva revised the Hippocratic Oath to provide modern doctors with a framework for safe and ethical practice.

Experiences during the Second World War bred new forms of welfare. In Britain, the administrative success of the Emergency Medical Service, which had been responsible for co-ordinating

DECLARATION OF GENEVA, 1948

At the time of being admitted as a member of the medical profession:

I solemnly pledge to consecrate my life to the service of humanity;

I will give to my teachers the respect and gratitude that is their due;

I will practise my profession with conscience and dignity;

The health of my patient will be my first consideration;

I will respect the secrets that are confided in me, even after the patient has died;

I will maintain by all the means in my power, the honour and the noble traditions of the medical profession;

My colleagues will be my sisters and brothers;

I will not permit considerations of age, disease or disability, creed, ethnic origin, gender, nationality, political affiliation, race, sexual orientation, social standing or any other factor to intervene between my duty and my patient;

I will maintain the utmost respect for human life;

I will not use my medical knowledge to violate human rights and civil liberties, even under threat;

I make these promises solemnly, freely and upon my honour.

regional health services during the war, provided a blueprint for the creation of a National Health Service (NHS) almost immediately after the war. Although women's health may have deteriorated during the war as the result of rationing, new antenatal and infant welfare centres were created immediately after the war, to improve maternal and child health. Renewed attention was also paid to global health initiatives. After the First World War, the League of Nations had been established to preserve international peace and protect global health. Following the Second World War, the WHO was founded to co-ordinate international efforts to control or eradicate infectious diseases, improve public health, and promote 'physical, mental and social well-being' across the world. One of the most striking successes of the WHO was against smallpox; the last recorded case of the disease was in Somalia in 1977. Three years later, the World Health Assembly officially announced that smallpox had been eradicated.

Work and welfare

All societies have recognized the importance of safeguarding the health and welfare of their citizens. In the distant past, religious orders, charities and state authorities provided poor relief, medical support and asylum to impoverished, ill and marginalized members of society. State and private interventions were often designed either to maintain the productivity of workers and soldiers or to preserve social order by reducing levels of poverty, vagrancy, crime and violence. Egyptian and Roman employers provided surgery to treat injuries and initiated precautionary measures to prevent the spread of infectious diseases amongst workmen building pyramids, cities and transport networks. As well as providing sanctuary and support to the sick and indigent, medieval hospitals and monasteries protected the rest of society from the threat of contagion and corruption. The English Poor Laws, introduced in the sixteenth and seventeenth centuries, were designed not only to improve the health but also to reduce the economic burden of the elderly, poor and infirm.

Late nineteenth- and early twentieth-century approaches to welfare were similarly driven by mixed motives. During the middle decades of the nineteenth century, sanitary reform and philanthropic support for hospitals were not only intended to boost public health but also to enhance national and imperial strength and promote industrial productivity. Towards the end of the century, on both sides of the Atlantic, those working in particular trades, and their families, could obtain health insurance through friendly societies, which in return for regular payments into the fund provided money for medical assistance as well as financial provision for funerals. Although there were suspicions that some infants were murdered in order to get the insurance payment from the 'burial clubs', these societies provided health care and social support, particularly to workers who could not otherwise afford it. Local friendly societies tended to cover only workers in urban centres but some of the societies expanded and evolved into large insurance companies.

More formal state mechanisms for promoting both individual and collective welfare were introduced in many countries from the late nineteenth century. In 1883, the German chancellor Otto von Bismarck (1815–98) passed legislation aimed at providing workers with sickness insurance. If they became ill, workers were provided with income and medical care financed by compulsory contributions from both employers and employees. Bismarck's model of health insurance was a blueprint for the provision of insurance-based welfare services across Europe and beyond, notably in Denmark, Belgium, Norway, Sweden, France, Japan and post-revolutionary Russia. A different form of welfare system was established in the United States. Attempts to introduce work-related insurance schemes based on the European model failed during the early decades of the twentieth century, largely because of political and professional opposition. Instead various voluntary schemes, such as the Blue Cross and Blue Shield prepayment plans, and private insurance companies, dominated provision. Not until the introduction of Medicare for the elderly and Medicaid for the poor in the 1960s and the health-care reforms introduced by President Barack Obama (b. 1961) in 2010 were these largely neglected sectors of American society provided with more reliable access to medical care.

In Britain, the Liberal government introduced National Health Insurance in 1911, following a visit made by the Chancellor of the Exchequer, David Lloyd George (1863–1945), to Germany three years earlier. Under the British scheme, weekly financial contributions (fourpence for men and threepence for women) to the fund made by employers and their employees were used to provide workers with cash benefits and medical care in the event of sickness and disability. Working women, and the wives of working men, received a maternity benefit of thirty shillings ($£1.50$). Hospital care was not included, except in the case of tuberculosis, which remained a major cause of death and disability. Although the National Health Insurance system improved access to health care, generated funds to establish the Medical Research Council in 1913, and provided doctors with regular work as 'panel doctors', it was intended primarily to reduce pauperism and facilitate workers' return to work. State intervention in health and welfare across the world in the early twentieth century should therefore be seen as part of wider social and economic reforms aimed at improving national efficiency and military strength, a notion encapsulated in the Japanese slogan 'Healthy people, Healthy Soldiers'.

Political expediency and military demand also shaped the development of child and maternal welfare provisions in many countries. Maternal mortality rates remained high in the inter-war years, fuelling concerns about rearing children who could become fit and able workers and soldiers. In some countries, infections and haemorrhage led to as many as 70 or 80 women dying for every 10,000 births. In Britain, approximately 1 in 200 women died during labour, a situation highlighted by Queen Mary in 1928, which led to the creation of a new charity, the National Birthday Trust Fund, that year. In the United States, where maternal mortality rates were among the highest in the world, approximately 20,000 women died each year during or after childbirth. Maternal deaths may have been the result of poor education in obstetric care, limited services for pregnant women and increasing levels of infection following illegal 'back-street' abortions. Closer monitoring of pregnancy and better training for midwives and doctors helped to reduce maternal deaths in

Britain during the 1930s but falling mortality rates were also the result of improved nutrition, the availability of antibiotics and the use of blood transfusions after the Second World War.

Inter-war welfare measures were hampered by global economic recession. Around the world, rising populations, increasing demands for health-care provision and pressures to purchase X-ray machines, modernize operating theatres and expand hospital laboratories led to substantial financial constraints and growing public and political dissatisfaction with failing medical services. By the outbreak of the Second World War, neither state nor charitable organizations could cope. In Britain, the response to this welfare crisis was the creation of the NHS in 1948. Funded by taxation, the NHS provided universal healthcare coverage that was free at the point of access and included treatment by both hospital consultants and general practitioners. Although a similar system had been introduced in New Zealand between 1938 and 1941, the British NHS was regarded as a pioneering initiative that was 'the envy of the world'. Other countries responded to economic and health-care crises in different ways. In Europe and Japan, social insurance schemes were extended, while health and welfare provision in the United States continued to be funded by the voluntary and private sectors.

Although post-war welfare initiatives helped to improve medical care, particularly for those previously unable to afford it, they were not without problems or controversy. In Britain, hopes that the NHS would improve the nation's health and lead to a reduction in health spending were confounded by the spiralling cost of new drugs and equipment and by patients' increasing demands for up-to-date treatments. Within a few years, the NHS was no longer able to supply prescriptions, spectacles or dental treatment entirely free of charge. Deaths from infectious diseases certainly declined but expanding state welfare services failed either to prevent rising levels of chronic conditions, such as cancer, heart disease, peptic ulcers, arthritis, diabetes and mental illness, or to close the gap between social classes. Statistics suggest that the upper classes were the major beneficiaries of welfare reforms and that lower income groups, ethnic minorities and migrants continued to bear the burden of chronic and work-related disease throughout the twentieth century. Globally, developing countries lagged behind Western countries in terms of mortality rates and life expectancy. In the mid-1960s, life expectancy at birth was less than fifty years in Guatemala, approximately sixty in Colombia, but over seventy in Norway, Denmark and Italy.

It is not yet clear what impact different welfare systems had on patterns of illness and death. Did the precise form of welfare provision matter for patients and their families? Has the NHS been better than insurance-based schemes or than the American system of private healthcare provision? Although life expectancy rates rose and infant mortality rates fell in all countries during the second half of the twentieth century, there is some historical evidence to indicate that countries operating social insurance schemes fared better than those with either private or British-style national health systems. However, the situation is complex, since other factors, such as education, social reform, environmental changes, the growth of occupational health services, and trends in smoking, exercise and diet played a significant part in shaping patterns of health and sickness. In many countries, rising rates of chronic disease stimulated campaigns for better public health education to prevent, or at least reduce, illnesses and deaths related to smoking, alcohol, poor diet, obesity and sex. During the closing decades of the twentieth century, campaigns were initiated in schools and disseminated through the media to highlight pressing health concerns, to educate young people about the risks associated with certain lifestyle choices and to boost the future welfare of modern populations.

Crises in the provision of welfare services in many countries after the Second World War, and concerns about the dehumanizing effects of scientific medicine, stimulated renewed interest in alternative or indigenous approaches to health care. In many parts of Eastern Europe, folk medicine persisted well into the twentieth century, particularly in rural communities. Traditional herbal remedies such as sassafras, St John's Wort and ginseng were revived, especially amongst European and American populations unable to afford orthodox Western medicines. Self-help and self-medication with over-the-counter preparations of medicines such as aspirin, paracetamol and ibuprofen remained a prominent form of medical care for many families. Rather than purchasing medicines from small pharmacists, patients in Britain and elsewhere increasingly bought cheaper medicines and spectacles from large retailers such as Boots, which had grown from a small herbalist store selling 'health for a shilling' in the mid-nineteenth century into an international pharmaceutical and cosmetics business by the middle of the twentieth century. Eastern and other forms of holistic medicine also became more popular in the West. Practitioners of acupuncture, yoga, Pilates, herbalism, chiropractic and homeopathy flourished as the affluent middle classes on both sides of the Atlantic turned away from conventional medicine in an effort to attain health and happiness by other means.

The commercialization of medicine was evident in the rising popularity of advice literature and products relating to personal and domestic hygiene. In the inter-war years, engaging in outdoor exercise and sport, the use of swimming baths and lidos and the adoption of healthy balanced diets were advocated as means of promoting health. After the Second World War, the expansion of media advertising and an increase in disposable income and leisure time encouraged families to purchase goods that would make their homes and bodies more beautiful. Soaps, detergents, vacuum cleaners, furniture polish and air fresheners were promoted on the grounds that they would improve the domestic environment by removing germs, dust and allergens from unhealthy homes. In some cases, such measures were linked to greater clinical awareness of the triggers of diseases such

Figure 18 Advertisement for toothpaste, 1940s (Wellcome Library, London)

as asthma and eczema. In others, they were driven by a general belief that cleanliness provided protection against the spread of infectious diseases such as colds and influenza.

At a more personal level, products to enhance appearance included pastes to whiten teeth (see Figure 18), creams to reduce skin blemishes and prevent ageing, shampoos and sprays to condition and shape hair, anti-perspirants and deodorants to eliminate body odour, varnishes to colour nails, waxes and creams to remove unwanted body hair, and new forms of safety and electric razors to ease shaving. Although most marketing efforts were aimed at women, who continued to be the major consumers of beauty products, towards the end of the twentieth century men began to purchase cosmetics and toiletries in greater quantities. During the same period, various forms of 'well-being' or 'wellness' therapy became popular, as both men and women visited private gyms, spas and saunas; paid for manicures, pedicures, facial treatments and massages; consulted counsellors and therapists to boost their capacity to cope with the stress of life; or resorted to cosmetic surgery to achieve the aesthetic form of the modern designer body. Even during periods of economic recession, the sales of health and beauty products continued to grow, creating an expanding market for the cosmetics industry not only in western Europe and the United States but also increasingly in Asia and South America.

The ability to purchase health and beauty was not always beneficial. One of the consequences of the increasing use of perfumes, dyes and cosmetics was a rise in cases of severe, and sometimes fatal, skin allergies. Rising levels of other forms of allergic diseases, such as asthma and hay fever, may have been triggered by better hygiene and by a reduction in childhood infections as the result of the use of antibiotics and vaccination. More extreme approaches to improving health and enhancing the body also led to medical problems during the twentieth century. Bodybuilders and fitness fanatics became addicted to anabolic steroids, leading to liver and heart disease. Breast implants ruptured, causing chronic and sometimes fatal side-effects. Botox injections, first introduced to remove facial wrinkles in the 1990s, were associated with infections and allergic reactions. Forms of surgery aimed at reducing weight, including liposuction or the insertion of gastric bands, carried risks. As patients became clients or customers of a wider range of practitioners, health and beauty became commodities available only to those who could afford them, accentuating inequalities in health across classes and regions. Driven by cultural obsessions with appearance and youth, the development and application of new techniques for manufacturing health and beauty thus generated new forms and patterns of disease.

The rise of biotechnology

We live in a technological world. Steam engines, the telegraph, automobiles, airplanes, space travel, computers, the Internet and mobile phones have successively transformed our environment, changing the nature of our relationships with each other and with ourselves. The benefits of such material shifts are evident. Modern populations have enjoyed greater opportunities for leisure, entertainment, education, enjoyment and self-fulfilment. But scientific progress has also created risks to health and happiness. The industrial and technological revolutions have spawned new forms of environmental degradation, pollution, occupational hazard, social disruption, terrorism and stress. As the nuclear accidents at Windscale, UK (1957), Three Mile Island, USA (1979) and Chernobyl, USSR (1986) demonstrate, the consequences of technological development can be devastating.

Similar patterns and paradoxes are evident in relation to advances in the biomedical sciences and biotechnology. During the twentieth century, diagnostic, therapeutic and managerial reforms generated new methods of treating disease, prolonging life and promoting health, and contributed to increased aspirations and expectations of individual health and well-being. At the same time, however, dramatic changes in the power of science and the delivery of health-care services produced new ethical dilemmas, particularly relating to the beginning and end of life, spawned new categories of medically induced (iatrogenic) disease, substantially altered the nature of the relationship between doctors and patients and exacerbated inequalities in health. In recent decades, medical progress has created its own forms of individual and social pathology.

Modern scientific medicine was conceived in the laboratories and clinics of western Europe and North America around the end of the nineteenth and beginning of the twentieth century, when a catalogue of discoveries and inventions transformed the practice of medicine. Some of the earliest developments involved measuring the function and structure of the heart. The first effective tool for determining blood pressure, the sphygmomanometer, was introduced in 1881 by the Austrian physiologist Samuel von Basch (1837–1905). During the early decades of the twentieth century, several prominent American physicians, including the neurosurgeon Harvey Cushing (1869–1939), who developed surgical techniques for successfully removing brain tumours, used the sphygmomanometer to measure systolic and diastolic pressure in the arteries. The ability to identify and monitor patients with high blood pressure, or hypertension, led to greater awareness of the consequences of not treating the condition. After the Second World War, studies showed that high blood pressure and high cholesterol levels greatly increased the risk of heart attacks and strokes, which by that time had become the leading causes of death in many countries.

Investigations into the function of the heart were aided by the development of the electrocardiogram (ECG), a tool that allowed doctors to chart the heart's electrical activity. Early attempts to record the human heartbeat had been made in the late nineteenth century and were developed further by the Scottish physician James Mackenzie (1853–1925), who used a modified kymograph, or polygraph, to trace and classify disturbances of the pulse. The first instrument to accurately measure the heart's electrical activity was introduced in 1903 by Willem Einthoven (1860–1927), who had been born in Indonesia but spent his working life as an ophthalmologist in Utrecht and later a professor of physiology in Leiden. Einthoven's apparatus was first used to study physiological processes in laboratory animals but was rapidly adopted by cardiologists for clinical purposes. The ECG facilitated the more precise diagnosis of various forms of heart disease, including angina, myocardial infarction (heart attack), atrial fibrillation and heart block, and the clearer correlation of cardiac function in the living heart with post-mortem findings. It also allowed scientists and clinicians to trace the effects of drugs such as digitalis on the rate and regularity of the heartbeat and to monitor the impact of exercise on patients with angina, a procedure that became known as a 'cardiac stress test'.

Perhaps the most striking scientific innovation in this period was the discovery and clinical application of X-rays. In November 1895, the German physicist Wilhelm Conrad Röntgen (1845–1923) noticed 'a new kind of ray' emanating from some paper soaked in barium platinocyanide. Röntgen's investigations revealed the ability of these electromagnetic rays to penetrate, or be obstructed by, a variety of materials and in December he produced the first-ever X-ray, of his wife's hand. Although the exposure time needed to produce reliable images was often up to thirty minutes, leading to an increased risk of skin damage and cancer, X-rays were quickly taken up by doctors around the world as a means of locating tumours and foreign bodies, demonstrating the damage caused by bullets and fragments of shrapnel during the Boer War and the First World War, and visualizing internal injuries to bones and joints (see Figure 19). By the 1920s, X-rays had become a pivotal tool in screening for tuberculosis, with public health campaigns encouraging people to be X-rayed, in order to avoid unnecessary deaths by detecting the disease as early as possible. Later in the century, low-energy X-rays became the standard means of screening women for early signs of breast cancer.

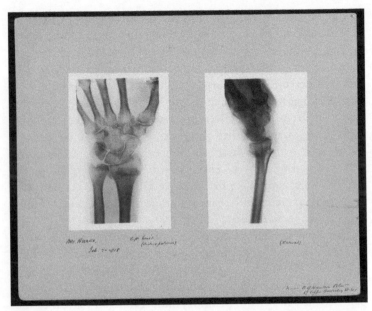

Figure 19 X-ray showing a fractured wrist, 1918 (Wellcome Library, London)

X-rays were not only of diagnostic importance. During his experiments, Röntgen had noticed the inflammation, ulceration and hair loss caused by prolonged exposure to X-radiation. Based on their ability to destroy cells and tissues, X-rays were first used to treat an American woman with breast cancer in 1896. Similar therapeutic procedures were introduced by European doctors to treat a variety of different cancers. Some approaches relied on X-rays while others used the radioactive properties of uranium and radium, which had been identified by Henri Becquerel (1852–1908). During the twentieth century, radiotherapy was improved by the development of more precisely directed rays and more carefully structured treatment programmes and was often used in conjunction with surgery and chemotherapy. The prevalence of malignant tumours amongst survivors of the atomic bombings of Hiroshima and Nagasaki in 1945 and the disaster at the Chernobyl Nuclear Power Plant in 1986 alerted the public and the medical profession to the dangers of radiation, leading to the introduction of safety measures during diagnostic and therapeutic procedures.

The limits and dangers of using X-rays as a diagnostic tool provided a stimulus for the development of safer and more sophisticated approaches to medical imaging. During the 1950s and 1960s, the radar and sonar techniques used to locate enemy submarines and aircraft during the Second World War were applied to the human body. Ultrasound waves were used by obstetricians to visualize the foetus and to detect placental anomalies. Similar techniques were used in the identification of heart disease, particularly abnormalities of the heart valves. Clearer representations of internal anatomy and pathology became possible in the 1970s, following the work of the British electrical engineer Godfrey Hounsfield (1919–2004), who was awarded the 1979 Nobel Prize for designing the first CT scanner, which used X-rays. Subsequent scanners used nuclear magnetic resonance or positron emission to generate high-resolution images capable of detecting physiological and biochemical abnormalities. In this way, doctors could see more clearly not only the macroscopic damage caused by cancers but also the less visible lesions found in neurological conditions such as multiple sclerosis.

Scientists' ability to examine the human body in even more detail was enhanced by the development of genetics. The term 'genetics' was introduced in 1906 by William Bateson (1861–1926) to describe the study of inheritance. Most early work on hereditary characteristics focused on plants but during the 1910s and 1920s scientists and clinicians began to explore the inheritance of resistance and susceptibility to diseases such as tuberculosis. The use of photomicrography to visualize chromosomes made it possible to identify the genetic anomalies responsible for conditions such as Down's syndrome, characterized by an extra chromosome 21, and the various intersex states, such as Klinefelter's and Turner's syndromes, that are characterized by an abnormal number of sex chromosomes. Elucidation of the biochemical structure of DNA by Francis Crick (1916–2004), Rosalind Franklin (1920–58), James D. Watson (born 1928), and Maurice Wilkins (1916–2004) in 1953 opened up new possibilities for detecting specific genetic defects, facilitating the diagnosis of haemophilia, sickle-cell anaemia, Huntington's chorea, cystic fibrosis and breast cancer, and leading to hopes that many of these conditions could in future be treated with gene therapy. In 1990, an international consortium of geneticists began the task of mapping the entire human genome, a project that was completed in 2003. Although tainted by its early association with eugenics and by subsequent anxieties about engineering designer babies, genetics has provided insights into disease that continue to affect human health.

The importance of more accurate diagnostic and screening tools was enhanced by the growing capacity of medical science to devise more effective treatments. Surgical advances during the later decades of the twentieth century included the ability to replace hip and knee joints, to transplant kidneys, lungs and hearts, to carry out less invasive 'keyhole' surgery, to repair damaged heart valves and bypass obstructed coronary arteries, to regulate the heartbeat by inserting a pacemaker, and to use microsurgical techniques in operations on the eyes, ears and brain. The success of these developments was dependent on early diagnosis, the use of dialysis and drugs to improve the pre-operative condition of patients, and the availability of blood transfusions and antibiotics to reduce post-operative complications. The likelihood of surviving life-threatening conditions was also linked to improvements in resuscitation and intensive care. For children, the development of the incubator served to halve infant mortality when it was first

LAPAROSCOPES AND ENDOSCOPES

The laparoscope was invented in the late nineteenth century to enable surgeons to see inside the abdomen without performing major surgery and was first used in the treatment of abdominal conditions in the early 1900s. By the late twentieth century, the use of cameras to visualize internal structures through the mouth, urethra or anus, a procedure referred to as endoscopy, had become commonplace. Laparoscopes and endoscopes are now used to detect and treat a wide range of conditions, including hernias, gastric ulcers, gallstones, bladder stones and torn cartilages and to tie the Fallopian tubes as a form of sterilization.

introduced in France in the late nineteenth century and provided the model for subsequent approaches to the intensive care of babies born prematurely.

Successful innovation was not confined to surgery. Diabetes had been described by ancient, medieval and early modern doctors, although its cause remained unknown. In the nineteenth century, the German physician Oskar Minkowski (1858–1931) suggested that diabetes was a disorder of the pancreas, but research failed to identify the hormone responsible and the administration of pancreatic extracts proved ineffective as a treatment. The management of diabetes was transformed in 1922, when the surgeon Frederick Banting (1891–1941), the biochemist James B. Collip (1892–1965), the physiologist J.J.R. Macleod (1876–1935) and a student assistant, Charles Best (1899–1978), isolated insulin from the pancreas of an ox and demonstrated its ability to lower blood sugar and restore lost weight and energy in patients with diabetes. The discovery of insulin, for which Banting and Macleod won the Nobel Prize in 1923, stimulated research into other hormone replacement therapies, most notably for diseases of the thyroid, pituitary and adrenal glands.

Vitamins, like hormones, could also be replaced to prevent or cure deficiency diseases. The clinical features of conditions such as scurvy (vitamin C deficiency), beriberi (vitamin B1) and rickets (vitamin D) had been known for many centuries and, in the case of scurvy, successfully treated. Studies in animals in the late nineteenth and early twentieth centuries revealed the role that nutrition played in many of these conditions and helped to isolate and synthesize the 'vital amines', subsequently referred to as vitamins, involved. A variety of vitamins was linked to specific diseases and vitamin preparations became a popular means of combating disease and promoting health. In addition, dietary advice became increasingly important, particularly in countries such as Japan, where beriberi, often referred to as 'Edo-disease' or 'Osaka-disease', was especially prevalent. Affecting particularly the nervous and cardiovascular systems, beriberi was discovered to be the result of a deficiency of thiamine (vitamin B1) and was more common in Japan because the thiamine-rich outer skin of rice, which was the staple food of many Japanese families, was removed during production. Associated with extreme poverty and urban living, rickets was revealed to be a disease caused by inadequate levels of vitamin D and insufficient exposure to sunlight. During the 1920s, scientists demonstrated that rickets could be prevented by the administration of natural or artificial light in conjunction with taking cod liver oil, which is rich in vitamin D, leading to a gradual reduction but not complete eradication of the condition.

Hopes that modern science could produce specific cures for specific diseases were also raised by the development of new vaccines for disabling conditions such as poliomyelitis, which was known to be transmitted by exposure to contaminated food and water. During the early decades of the twentieth century, summer epidemics of polio in many Western countries caused infantile paralysis, muscle wasting and sometimes death. Awareness of the effects of polio was increased after the American politician Franklin D. Roosevelt (1882–1945) contracted the condition in

1921. Paralysed and often wheelchair-bound, Roosevelt never-theless became president of the United States in 1933 and in 1938 established the March of Dimes Foundation to support polio research and education. During the 1950s and 1960s, the threat of paralysis and the fear of being treated in an 'iron lung' were diminished by the development of effective vaccines. The introduction in 1955 of a killed virus vaccine by the American virologist Jonas Salk (1914–95) led to a significant reduction in cases of polio, particularly in countries where vaccination was compulsory. However, the use of Salk's vaccine was limited by the fact that it was costly and dangerous. It was replaced by a live attenuated virus vaccine produced by the Russian-born scientist and physician Albert Sabin (1906–93) in the 1960s. The Sabin vaccine, which could be administered on a flavoured sugar lump to make it more palatable for children, initiated a further dramatic fall in the incidence of polio. By the early twenty-first century, outbreaks of polio occurred only occasionally in some parts of South-East Asia and Africa.

Before the Second World War, there were few reliable forms of birth control. For women wishing to prevent pregnancy, the practice of 'coitus interruptus' and the use of condoms, diaphragms and herbal preparations offered some protection against conception. Some women resorted to illegal abortions or, more rarely, infanticide to limit the number of their children. In some countries, such as China and India, cultural and economic preferences for sons over daughters may have led to the selective infanticide, abandonment or adoption of girls. These approaches were all problematic; not only were they often unsuccessful but they were also dangerous and, in some cases, criminal. For many religions, contraception was immoral, and in Western countries there were fears that the use of contraceptive measures by the middle classes would lead to degeneration of the population and a weakening of military and commercial power. The first birth control clinics were opened by Margaret Sanger (1879–1966) in America in 1916 and by Marie Stopes (1880–1958) in Britain in 1921 but they remained controversial.

The possibility of preventing pregnancy more effectively improved after the Second World War, when scientists in America began to test the ability of various steroid hormones to prevent conception. Their efforts were supported by the Planned Parenthood Federation of America and by influential philanthropists such as Katharine Dexter McCormick (1875–1967), who believed that a contraceptive pill would emancipate women and halt unsustainable population growth. In the 1950s, the American biologist Gregory Pincus (1903–67) and the gynaecologist and obstetrician John Rock (1890–1984) demonstrated the ability of hormone preparations to prevent pregnancy. By 1960, the first oral contraceptive pill, containing both an oestrogen and a progestogen, was marketed. Religious objections came from the Catholic Church and some Eastern cultures regarded 'the pill' as unnatural. There were also growing concerns about the health impact of the pill, in particular long-term use of oestrogen-based pills. Nevertheless, oral contraceptives became the most effective and most popular form of birth control throughout the modern world. More recently, the introduction of the 'morning-after pill', which makes it possible for women to prevent fertilization after unprotected sex, and early abortion pills such as RU-486, have re-energized disputes about the morality of contraception and abortion.

Debates about psychiatry were arguably even more polarized. During the eighteenth, nineteenth and early twentieth centuries, the mentally ill were either cared for at home or incarcerated in purpose-built asylums. Disruptive and violent patients might be restrained by chains or straitjackets, or sedated by bromides, opiates and barbiturates. From the late nineteenth century, manic patients were sometimes prescribed lithium to control episodes of agitation, delusion, aggression and excitability. More invasive and dramatic forms of therapy introduced during the middle decades of the twentieth century included electroconvulsive therapy (ECT), insulin shock therapy and lobotomy or leucotomy. This last procedure was devised in the 1930s by the Portuguese neurologist Egas Moniz (1874–1955), who argued that severe mental illness could be ameliorated by severing the connections between the frontal lobes and the rest of the brain. Although Moniz was awarded the Nobel Prize in 1949, lobotomy became highly controversial, partly because of the work of Walter Jackson Freeman (1895–1972), an American neurologist who performed over 3,000 lobotomies on unanaesthetized patients by inserting an instrument like an ice pick (a small, sharp, pointed tool) up through the eye sockets and moving it backwards and forwards to cut through the brain tissue. Patients were often left with severe personality disorders or died from cerebral haemorrhage, and Freeman was eventually banned from performing lobotomies.

Although ECT remained in use, brutal psychosurgical operations were increasingly rendered obsolete by the development of new drugs after the Second World War. Minor tranquillizers such as the benzodiazepines Valium and Librium, antipsychotic agents such as chlorpromazine (or Largactil), and antidepressants such as imipramine, amitriptyline and more recently fluoxetine (marketed as Prozac) were hailed as wonder drugs that would revolutionize the management of mental illness and allow patients to be treated more humanely in the community rather than in large, outdated, institutions. In many ways, these drugs fulfilled their early promise. Asylums closed, as patients appeared to recover from the delusions, hallucinations, insomnia, paranoia and suicidal thoughts associated with anxiety, depression, stress, schizophrenia and manic depression. The status of psychiatry was enhanced by its ability to demonstrate the rational scientific basis of psychiatric diagnosis and practice. Valium became so popular as a panacea for housewives battling domestic boredom and social isolation that it became known as 'Mother's little helper', a phrase immortalized in a 1966 song of that title by the Rolling Stones.

Medicating mental illness, however, became problematic. Psychiatric patients became addicted to drugs or suffered severe side-effects. In some cases, medication increased the risk of suicide, particularly among depressed young men. Contemporary preoccupations with chemical cures diverted political and professional attention from addressing the situational and socio-economic causes of mental illness, such as poverty, bereavement, marital problems and domestic violence. Criticisms of psychiatry were exacerbated by scepticism about the validity of conventional drug treatments, by fears that psychiatrists were labelling women as inherently unstable and irrational, and by concerns about the institutionalization and abuse of psychiatric inpatients, evocatively captured in a 1961 study of asylums by the sociologist Erving Goffman (1922–82), and by Ken Kesey's (1935–2001) 1962 novel, *One Flew Over the Cuckoo's Nest*. Spearheaded by writers such as Laing and Szasz, the anti-psychiatry movement condemned the tendency of medicine to extend the boundaries of mental illness to include normal emotions and behaviour and denounced the immorality of the pharmaceutical industry.

As the cases of birth control and psychiatry suggest, scientific and technological advances were not unequivocally beneficial. Many drugs and surgical interventions increased, rather than decreased, the incidence and severity of disease. The most infamous example is the case of thalidomide, which was introduced in West Germany during the late 1950s as a sedative and was also used to treat morning sickness in pregnant women. By 1961, thalidomide had been withdrawn, because of its capacity to produce birth defects in children born to mothers who had taken the drug during early pregnancy. During the 1960s and 1970s, it became apparent that the side-effects of many drugs were more severe and unmanageable than the symptoms of the diseases for which they were prescribed, leading to non-compliance and growing public interest in pursuing gentler and supposedly more natural pathways to health and happiness. Although Western medicine retained its position of professional dominance in many countries, alternative forms of medical practice, such as acupuncture, herbalism and homeopathy, continued to flourish as patients and their families struggled to find the most effective, but least disagreeable, forms of health care.

Many of the therapeutic innovations of the late twentieth century raised ethical and legal dilemmas. The shortage of organs for transplant led to debates about the best means of encouraging people to donate, with policies ranging from the introduction of donor cards, organ registers, improved co-ordination of transplant services and presumed-consent legislation. During the 1980s and 1990s, in countries such as Belgium, where consent to donate was presumed, the availability of organs for transplant increased dramatically but many countries, including Britain, continued to regard such legislation as politically contentious. The rise of transplant surgery initiated other debates, particularly about the precise diagnosis of death, the equitable distribution of scarce resources and the ethics of selling organs. Abortion and

ORGAN TRAFFICKING

In the 1980s, the shortage of organs available for transplant spawned a lucrative trade in organs. In 1989, it was discovered that four Turkish citizens had each sold one of their kidneys for cash. The scandal prompted the British government to pass the Human Organ Transplants Act 1989, which prohibited the sale of organs, on the basis that potential donors were vulnerable to coercion and exploitation. Although similar legislation prohibiting the sale of organs was passed in many other countries, including China and India, the continuing financial incentives have ensured that illegal organ trafficking persists.

new forms of contraception provoked disputes about the value of life and the legitimate limits of medical authority. Conversely, *in vitro* fertilization (IVF), embryo research and cloning technologies raised the spectre that medical science might create new, and potentially dangerous, life forms.

Technological solutions to modern health dilemmas substantially increased the power of the state to intervene in private lives. One clear example of this apparent encroachment on individual liberty is evident in debates about how best to reduce the burden of disease caused by smoking. In the 1950s, the British epidemiologists Austin Bradford Hill (1897–1991) and Richard Doll (1912–2005) demonstrated a causal link between cigarette smoking and lung cancer. Although their findings were initially rejected, particularly by the tobacco industry, the association between smoking, cancer and heart disease was subsequently acknowledged by scientists, clinicians and the public. In spite of the evidence, however, men and women continued to smoke, contributing to rising levels of chronic disease. Policy responses included health education initiatives encouraging smokers to take responsibility for their health, the development of new treatments, increasing tax on cigarettes and the introduction of compulsory restrictions on smoking in public. Although effective in reducing damage caused by passive smoking, such legislation has been criticized as interfering with individual liberty. As in earlier centuries, modern medicine has had to tread a fine line between protecting the freedom of individuals and preserving the health of the nation.

The limits of medicine?

In the early 1980s, a new plague emerged to shatter the confident post-war assertions that modern medicine would soon overcome human disease. Outbreaks of opportunistic infections and rare forms of cancer among members of the gay community in America heralded a revival of fears about the resurgence of contagious diseases for which there appeared to be no treatment. As with previous epidemics of sexually transmitted diseases, what became known as acquired immunodeficiency syndrome (AIDS) provoked a moral outrage most notably evident in the public condemnation of homosexuality. However, AIDS was rapidly identified as a viral condition that affected both heterosexuals and homosexuals, spread by sexual contact, sharing infected syringes, and blood transfusions. By 1986, approximately 21,000 cases of AIDS had been reported worldwide, with a mortality rate of nearly 100%. Public anxiety and the activism of gay communities prompted greater scientific efforts to develop anti-viral treatments and to boost immunological resistance but by the late 1980s five to ten million people were thought to have been infected with the human immunodeficiency virus (HIV) that caused AIDS. Policy responses varied but increasingly focused on educating people about the dangers of unprotected sex and encouraging the use of condoms. Although these policies, and the creation of therapeutic medicine cocktails, have reduced the spread and impact of AIDS, the disease remains a major killer in many underdeveloped regions of the world.

AIDS was not the only cause for medical concern. During the closing decades of the twentieth century, other scourges threatened the health and well-being of modern human and animal populations. Some of these were relatively new: bird flu, severe acute respiratory syndrome (SARS), foot-and-mouth disease, chlamydia, bacterial resistance to antibiotics, and the rising prevalence of severe allergic reactions were health hazards that appeared to be generated by modern patterns of travel, sexual activity and hygiene. Other conditions were more familiar manifestations of older diseases exacerbated by poor diet, insufficient exercise and social deprivation: rising levels of obesity; the return of tuberculosis, mumps and measles; dramatic increases in mental illness, autism and hyperactivity amongst children, and dementia in the elderly, served to remind doctors, patients and policymakers that the battle against disease was by no means won. In spite of substantial advances in medical knowledge in recent decades, the eradication of disease has proved to be unattainable.

The appearance of new diseases and the return of old plagues have neither entirely demoralized modern populations nor led to the abandonment of scientific research and biomedical innovation. Certainly, some doctors and patients have expressed concerns about the inability of scientific medicine to address many chronic health problems and called for a revival of holistic approaches to health care. However, rather than deflating biomedical aspirations, modern patterns of health and disease have stimulated renewed efforts to identify causes, develop treatments, educate patients and improve the management and delivery of health-care services. Scientific research and health policies have often been constrained by economic recession, political vacillation and cultural obstacles, but formulating strategies for combating disease and promoting health remains a priority for national and international health authorities. Given the remarkable successes achieved over the last century or so, it would be premature to suggest that we have reached the limits of medicine.

Conclusion

We can indeed expect a 'new chapter in the history of medicine'
but the chapter is likely to be as full of diseases as its predecessors;
the diseases will only be different from those of the past.

René Dubos, *The Dreams of Reason*, 1961

Medicine evolves. Over two thousand years ago, Hippocrates
suggested that epilepsy was not caused by divine or demonic
influences, as many people believed, but by the accumulation of
phlegm in the brain. According to Hippocrates, epilepsy was best
treated, like other diseases, by restoring humoral balance rather
than by employing spells and sacrifices. In the early twenty-first
century, such theories of the causes of disease seem outmoded.
We regard epilepsy as a neurological condition characterized by
sudden and excessive electrical activity in parts of the central
nervous system. Seizures of various types can be triggered by
fevers, flashing lights, trauma, drugs and tumours. Diagnosed
with the aid of electroencephalograms and MRI scans, epilepsy
is currently managed by avoiding known triggers and taking
anticonvulsant medication such as carbamazepine, phenytoin,
sodium valproate, gabapentin and lamotrigine.

Medicine also stays the same. Although the precise definitions
of many categories of disease have changed, conditions that were
identified and named by ancient, medieval and early modern
medical authors, such as epilepsy, diabetes, asthma, leprosy and
melancholy, remain recognizable. Modern patients with asthma
describe their symptoms similarly to the description of breathlessness provided by the Stoic philosopher Seneca during the first

century CE. In addition, neither the ambition of doctors to reduce disease and promote health nor the boundaries of acceptable practice have altered significantly over the centuries. Many of the central tenets of the Hippocratic Oath, such as the promise not to harm patients and a commitment to confidentiality, continue to provide a framework for the regulation of medical practice in the modern world. The history of medicine is characterized by elements of continuity as well as change.

What can we learn from the history of medicine? In the first instance, historical studies reveal the manner in which medical knowledge is rarely either uncontested or stable. Competing forms of knowledge and practice have always existed side by side, irrespective of period or location. Patients in every age have been able to consult a wide range of healers and embrace many different forms of health care. Specific approaches to health and disease have dominated particular eras and civilizations but they have never completely excluded alternative therapeutic strategies. What appears to be orthodox medicine in one period or place can be marginalized in another, as scientific investigations and technological developments serve to transform medical theory and practice and to redefine the boundary between health and disease. Even modern medicine, which we presume to offer better health care than in the past, remains fragile and will surely be surpassed.

Second, history suggests that patterns of disease and shifting medical understandings of health have both been shaped by historical, social, cultural and political forces. Just as the appearance, transmission and consequences of disease have been determined by poverty, migration, trade and military conflict, changing formulations of the body and rival approaches to disease prevention and control have been moulded by the sociopolitical context. Eighteenth-century divisions between orthodox and alternative practitioners, nineteenth-century disputes between 'miasmatists' and 'contagionists', and twentieth-century reservations about the dehumanizing effects of biotechnology owed as much to contemporary social and cultural values as they did to the science that underpinned practice. The form and content of scientific and clinical knowledge, as well as the structure of health-care services, are always contingent on prevailing moral standards, economic constraints and the balance of political power.

Third, there have always been drawbacks to medical progress. Although Renaissance and early modern understandings of the body in health and disease led, perhaps slowly and indirectly, to improvements in surgery, they also initiated a macabre trade in bodies and body parts. The formulation of chemical approaches to disease and the exchange of remedies across national borders in the seventeenth and eighteenth centuries may well have expanded the options available to the sick and dying but they also generated unwelcome side-effects and made new poisons available to aspiring murderers. A growing Enlightenment reliance on reason and science stigmatized the insane and discriminated against women. In the late nineteenth and early twentieth centuries, one of the consequences of the emergence of genetics and eugenics was the formulation of repressive policies to segregate, sterilize and eliminate those considered to be unfit. The double-edged nature of advances in medicine has persisted into the twenty-first century. The increasing capacity of Western biomedicine to create or prolong life through artificial means has resulted in ethical dilemmas that continue to challenge modern societies.

Finally, history teaches us that there can be no victory over disease and death. Although we may be able to prevent or alleviate pain and suffering, we can never entirely escape from the clutches of sickness and disability. As the French-born microbiologist René Dubos (1901-82) pointed out in the 1960s, during a golden age of medicine when doctors and their patients were dreaming of a medical Utopia, new diseases always emerge to replace those that have supposedly been conquered. Social deprivation, famine, warfare and pollution, as well as an inexorable rise in modern populations, will continue to threaten global health. This forecast should not lead to therapeutic pessimism and the rejection of biomedicine but to renewed efforts to improve diagnostic and therapeutic procedures, to promote social and welfare reform and to discover alternative, perhaps more individual and creative, pathways to health and happiness.

How can medical history and the medical humanities contribute to this broader humanitarian goal of improving health and well-being? Research in the humanities reminds us that bioscientific models of disease, according to which illness is merely the product of disordered chemical and physiological mechanisms, are limited. While biomedicine has certainly produced startling advances in understanding and treatment, it has generally failed to acknowledge or address the historical, social and cultural forces that determine the origins, manifestations, experiences and meanings of disease. The art of medicine requires attention to bodies and cultures, to mechanisms and meanings, to health and disease, and to the interactions between past and present. To achieve more balanced and fertile accounts of health and disease, we must collapse simple distinctions between science and the humanities and foster a spirit of enquiry that integrates the insights of both. In this way, we may be able to cope more effectively with the diseases and despair that will surely afflict us in the future.

Timeline

3rd millennium BCE	Huangdi, *The Yellow Emperor's Classic of Internal Medicine*, a treatise on the causes and symptoms of disease

3rd millennium BCE — Huangdi, *The Yellow Emperor's Classic of Internal Medicine*, a treatise on the causes and symptoms of disease

Origins of Indian Vedic or āyurvedic medicine, subsequently transcribed in the *Caraka Samhitā* and *Súsruta Samhitā*

2nd millennium BCE — Egyptian papyri, detailing signs and symptoms of disease, surgical procedures and many herbal treatments

c. 490 BCE — Birth of Alcmaeon of Croton, who performed dissections and emphasized the importance of maintaining bodily balance

c. 460 BCE — Birth of Hippocrates, sometimes referred to as the 'father of medicine', who advocated natural explanations of disease and humoral theory

25 BCE — Birth of Celsus, Roman author of *De medicina*

c. 40–90 CE — Dioscorides, author of *De materia medica*, the earliest Western encyclopaedia of herbal remedies

129–210 CE — Galen, influential Greek physician, who developed, refined and transmitted Hippocratic medicine

354 CE — Birth of St Augustine, who emphasized the healing power of Christ

5th–6th centuries CE	Justinian plague killed 1,000 people each day, contributing to the fall of the Roman Empire and the rise of Byzantium
6th–10th centuries CE	Sui (581–618) and Tang (618–907) periods, during which Chinese medicine was strongly shaped by Daoism and Buddhism
865	Birth of al-Rāzī, author of *The Comprehensive Book of Medicine*
936	Birth of al-Zahrāwi, who performed surgical procedures such as cautery, suturing and the treatment of fractures and bladder stones
980	Birth of Ibn Sīnā, author of the *Canon of Medicine*
1098–1179	Hildegard of Bingen, prominent theologian and healer
11th century	ūnānī tibb formed from the merger of indigenous Indian, Greek and Islamic medicine
	Gilles de Corbeil (1165–1213) introduced the matula to examine urine
12th–13th centuries	Early hospitals founded, including St. Bartholomew's (1123), St Thomas's (1215) and Bethlem (1247) in London
1126	Birth of Ibn Rušd, author of an encyclopaedia on general medicine
1138–1204	Maimonides, author of the first treatise on asthma
1340s	Spread of the 'Black Death', which killed 20 million Europeans
1353	*Decameron*, an account of the plague written by Giovanni Boccaccio (1313–75)

1363	Contemporary description of the devastation caused by the plague written by Guy de Chauliac (1300–68)
1368	English Guild of Surgeons founded
1452–1519	Leonardo da Vinci; dissected bodies and produced anatomical illustrations
1462	Guild of the Barbers of London founded
1466–1536	Desiderius Erasmus, Catholic priest and humanist scholar, who promoted ethical standards within medicine and advocated legislation to improve public health
1490s	Appearance of syphilis in Europe
1493–1541	Paracelsus, a Swiss doctor, who rejected ancient authorities, preferring to observe nature in order to gain information about disease; promoted a chemical account of the body
c. 1500–74	Bartolomeo Eustachi, who described the Eustachian tube connecting the middle ear with the nasopharynx
1510–90	Ambroise Paré, renowned French army surgeon, who developed new techniques for treating wounds and demonstrated that the bezoar stone did not cure patients who had been poisoned
1523–62	Gabriele Falloppio, who described the Fallopian tube connecting the ovaries to the uterus
1536–1614	Felix Platter, a Swiss physician, who described the symptoms of mania and melancholy

1540	Company of Barber–Surgeons established in England
1543	Publication of Andreas Vesalius's (1514–64) *De humani corporis fabrica*, the most influential Renaissance text on anatomy
	Nicolaus Copernicus proposed a heliocentric model of the universe in *De revolutionibus orbium coelestium*
1560–1631	Pierre Chamberlen, inventor of the obstetric forceps
1579–1644	Joan Baptista van Helmont, Flemish nobleman, who endorsed the chemical theories of Paracelsus and adopted the 'doctrine of signatures'
1596–1650	René Descartes, who regarded the body as a machine
1621	Publication of Robert Burton's (1577–1640) *The Anatomy of Melancholy*, the first English treatise on madness
1621–75	Thomas Willis, English physician, who described the circle of arteries supplying blood to the brain, now known as the 'circle of Willis'
1624–89	Thomas Sydenham, sometimes referred to as the 'English Hippocrates'
1628	William Harvey published *De motu cordis*, demonstrating the importance of the heart in pumping blood around the body and the action of the valves in the veins
1652	Nicholas Culpeper published *The English Physician*, which included advice on medicinal herbs

1665	Robert Hooke published *Micrographia*, which demonstrated the power of the microscope to reveal the fine structure of cells
1668–1738	Hermann Boerhaave, leading Enlightenment medical theorist
1671	Publication of Jane Sharp's *The Midwives Book*, an early manual of midwifery
1718–83	William Hunter, one of the first man-midwives, who set up a private anatomy school in London
1720s	Lady Mary Wortley Montagu (1689–1762) introduced smallpox inoculation into Britain from Turkey
1720–91	Lán Ông, Vietnamese doctor
1728–93	John Hunter, leading British surgeon
1734 1815	Anton Mesmer, founder of mesmerism
1741	Foundling Hospital founded in London
1745	British Company of Surgeons formed
1745–1826	Philippe Pinel, who introduced psychological therapy for the insane
1749	Founding of Lying-In Hospital for Married Women, in London
1753	James Lind (1716-94) demonstrated the use of oranges and lemons to prevent scurvy in sailors
1755–1843	Samuel Hahnemann, founder of homeopathy
1758–1828	Franz Joseph Gall, co-founder of phrenology
1760–1835	Hanaoka Seishū, Japanese surgeon, who used herbs to induce anaesthesia
1761	Leopold Auenbrugger (1722–1809), Viennese physician, introduced percussion of the chest

	Publication of Giovanni Battista Morgagni's (1682–1781)*De sedibus et causis morborum*, translated into English in 1769
1766–1832	JC Spurzheim, co-founder of phrenology
1766–1842	Dominique Jean Larrey, French surgeon, introduced stretcher-bearers, mobile field hospitals and early ambulances on to the battlefield
1770	John Gregory (1724–73) published *Observations on the Duties and Offices of a Physician*
1785	William Withering (1741–99) published an account of the medical uses of foxglove (*digitalis*)
1793	Matthew Baillie (1761–1823), *Morbid Anatomy*
1793–1860	Thomas Addison, who described adrenal insufficiency, now known as Addison's disease
1794	Thomas Percival (1740–1804), published *Medical Ethics* (revised and expanded in 1803)
1795–1862	Thomas Wakley, founding editor of the *Lancet*
1796	Edward Jenner (1749–1823) introduced vaccination (using cowpox) for the prevention of smallpox
	Opening of the York Retreat, where 'moral treatment' for the insane was pioneered
1798	Lying-In Hospital of the City of New York founded
1798–1866	Thomas Hodgkin, who described a form of lymphoma

1800	Royal College of Surgeons in London founded
1805–81	Mary Jane Seacole, Jamaican nursing reformer
1807	Smallpox vaccination made compulsory in Germany
1808–77	Sir William Fergusson, Scottish surgeon who pioneered operative repair of cleft lips and palates
1809–85	Friedrich Gustav Jacob Henle, who described the 'loop of Henle' in the kidney
1809–94	Oliver Wendell Holmes, American doctor, who advocated washing with chlorinated water to prevent infections
1812–73	Isaac Baker Brown, infamous English surgeon, who performed clitoridectomies to curb female sexual desire
1815	Apothecaries Act introduced better regulation of apothecaries in England and Wales
1816–88	Carl Zeiss, German mechanic, manufacturer of microscopes
1817	James Parkinson (1755–1824) described the 'shaking palsy', now known as Parkinson's disease
1819	Stethoscope invented by René Théophile Hyacinthe Laennec (1781–1826)
1820–1910	Florence Nightingale, who reformed nursing practice during the Crimean War
1821–1910	Elizabeth Blackwell, one of the first women to be admitted to medical school
1822–95	Louis Pasteur, who first demonstrated the use of vaccination to treat rabies, anthrax and

	chicken cholera and developed a method of removing germs from milk, beer and wine by heating ('pasteurisation')
1822–1911	Francis Galton, British statistician, who developed the field of eugenics
1825–82	Joseph T. Clover, English doctor who pioneered the use of anaesthetics during surgery
1829	Execution of William Burke, who, with William Hare, had murdered sixteen people and sold the bodies to surgeons
1832	Passage of the Cholera Prevention Act and the Anatomy Act in Britain
1836–1917	Elizabeth Garrett Anderson, one of the first women to be admitted to medical school
1839–1910	Akashi Hiroakira, Japanese physician who integrated traditional healing practices and Western medicine
1840s	Public baths and wash-houses built in Britain to promote cleanliness
1840–1912	Sophia Jex-Blake, who qualified in medicine in Switzerland and founded medical schools in London and Edinburgh
1842	*Report on the Sanitary Condition of the Labouring Population of Great Britain* by Edwin Chadwick (1800–90)
	William E Clarke (1819–98) used ether as an anaesthetic to extract a tooth
1842–1906	Mary Putnam Jacobi, one of the first American women to be admitted to medical school

1843–1910	Robert Koch, German physician, who identified the organisms responsible for tuberculosis (1882) and cholera (1884)
1844	Introduction of vulcanized rubber leading to safer condoms
	Horace Wells (1815–48) used nitrous oxide as an anaesthetic
1846	William TG Morton (1819–68) administered ether to remove a neck tumour
1847	Ignaz Phillipp Semmelweis (1818–65) reduced mortality during childbirth by ordering doctors and midwives to wash their hands in chlorinated lime
	James Young Simpson (1811–70) discovered the anaesthetic properties of chloroform
1848	Public Health Act, Britain
1853	Smallpox vaccination made compulsory in Britain
	John Snow (1813–58) administered ether to Queen Victoria during the birth of Prince Leopold
1853–1925	Scottish physician James Mackenzie, who used the kymograph to trace the human pulse
1854	John Snow demonstrated the role of contaminated water in transmitting cholera
1854–6	Crimean War, amputations performed under anaesthesia
1854–1915	Paul Ehrlich, German scientist, developed 'salvarsan' for the treatment of syphilis

1858	Medical Act, which created the British General Medical Council and set the educational standards of the profession
1858–1931	Oskar Minkowski suggested that diabetes was a disorder of the pancreas
1860s	Women gradually admitted to European medical schools
1865	Joseph Lister (1827–1912) introduced carbolic acid as an antiseptic
1880s–90s	Influenza pandemics
1881	Tianjin Medical School opened in China
	Sphygmomanometer introduced to measure blood pressure by Samuel von Basch (1837–1905)
1882	Tuberculosis organism discovered by Robert Koch
	Takaki Kanehiro (1849–1920), a Japanese naval surgeon, prevented beriberi in sailors by substituting barley for rice
1882–1960	Sir Harold Gillies, pioneering plastic surgeon
1883	Sickness insurance introduced by the German chancellor Otto von Bismarck (1815–98)
1884	Cholera organism discovered by German bacteriologist Robert Koch
1887	Pasteur Institute, Paris, founded
1890s	Introduction of Aspirin by Friedrich Bayer and Company
1891	Lister Institute, London, founded
	Emil von Behring (1854–1917) and Shibasaburō Kitasatō (1856–1931)

	demonstrated the efficacy of diphtheria antitoxin
1895	Wilhelm Conrad Röntgen (1845–1923) demonstrated X-rays
1896	First use of X-rays to treat cancer
1899–1902	X-rays used to locate bullets during the Boer War
1900–60	Sir Archibald McIndoe, pioneer in plastic surgery
1903	Electrocardiogram (ECG) introduced by Willem Einthoven (1860–1927)
1905	Alexis Carrel (1873–1944) performed first heart transplant in a dog
1906	First kidney transplant in a human performed by Mathieu Jaboulay (1860–1913)
	William Bateson (1861–1926) coined the term 'genetics'
1911	British National Insurance Act
1914–18	First World War, during which new treatments for wounds, advance dressing stations, fracture clinics, mobile bacteriological units, and plastic surgery were developed
1916	Margaret Sanger (1879–1966) introduced birth control clinics in America
1918	Pandemic Spanish influenza killed 20 million people world-wide
	Maternal and Child Welfare Act, Britain
1919–2004	Godfrey Hounsfield, who developed the CT scanner

1921	Bacille Calmette-Guérin (BCG) vaccine introduced for tuberculosis
	Marie Stopes (1880–1958) set up the first birth control clinic in Britain
1922	Discovery of insulin by Frederick Banting, James B Collip, JJR Macleod and Charles Best
1928	Discovery of penicillin by Alexander Fleming (1881–1955)
1930s	Introduction of ECT, insulin shock therapy and lobotomy for patients with mental illness
1932	American Tuskegee Syphilis Study
1938–41	Introduction of a national health system in New Zealand
1939–45	Second World War, during which Howard Florey, Ernst Chain and Norman Heatley manufactured penicillin for the Allied troops
1943	Kidney dialysis machine first used by Willem Kolff (1911–2009)
1947	Nuremberg Code
1948	World Health Organization founded
	Declaration of Geneva
	National Health Service established in Britain
1950s	Development of contraceptive pill by Gregory Pincus (1903–67) and John Rock (1890–1984)
	Austin Bradford Hill (1897–1991) and Richard Doll (1912–2005) identified a link between smoking and lung cancer
1953	Francis Crick, James D. Watson, Maurice Wilkins and Rosalind Franklin revealed the biochemical structure of DNA

1955	Jonas Salk (1914–95) introduced a killed polio vaccine
1960s	Introduction of a live polio vaccine by Albert Sabin (1906–93)
	Benzodiazepines, such as Valium and Librium, introduced for anxiety, depression and stress
	Introduction of Medicaid and Medicare in America
1961	Thalidomide withdrawn because of birth defects in the children of mothers who had taken the drug during pregnancy
1964	Declaration of Helsinki
1967	First heart transplant performed by Christiaan Barnard (1922–2001)
1977	WHO officially declared the eradication of smallpox
1978	Birth of Louise Brown, the first 'test-tube' baby
1980s	AIDS first recognised in America
	Hepatitis B vaccine
1990s	Stem cell therapy developed
1990–2003	Human Genome Project
1996	Hepatitis A vaccine
2003	SARS (severe acute respiratory syndrome) appeared
2009	Swine flu (H1N1) identified

Further reading

The following reading list comprises a selection of books on the history of medicine, ranging from scholarly studies to accessible overviews and incorporating a variety of historical perspectives.

Introduction

John Burnham, *What is Medical History?* (Oxford, Polity Press, 2005).

William Bynum, *The History of Medicine: A Very Short Introduction* (Oxford, Oxford University Press, 2008).

William Bynum and Helen Bynum (eds.), *Great Discoveries in Medicine* (London, Thames & Hudson, 2011).

W.F. Bynum and Roy Porter (eds.), *Companion Encyclopaedia of the History of Medicine Vols. 1 and 2*, (London, Routledge, 1993).

Jacalyn Duffin, *History of Medicine: A Scandalously Short Introduction* (Toronto, University of Toronto Press, 1999).

Frank Huisman and John Harley Warner (eds.), *Locating Medical History: The Stories and Their Meanings* (Baltimore, Johns Hopkins University Press, 2006).

Mark Jackson (ed.), *The Oxford Handbook of the History of Medicine* (Oxford, Oxford University Press, 2011).

Irvine Loudon (ed.), *Western Medicine: An Illustrated History* (Oxford, Oxford University Press, 1997).

Roy Porter, *The Greatest Benefit to Mankind: A Medical History of Humanity from Antiquity to the Present* (London, HarperCollins, 1997).

Roy Porter, *Blood and Guts: A Short History of Medicine* (London, Penguin, 2003).

Paul Strathern, *A Brief History of Medicine* (London, Robinson, 2005).

Keir Waddington, *An Introduction to the Social History of Medicine* (Basingstoke, Palgrave Macmillan, 2011).

David Wootton, *Bad Medicine: Doctors Doing Harm Since Hippocrates* (Oxford, Oxford University Press, 2006).

1

Jan van Alphen and Anthony Aris (eds.), *Oriental Medicine: An Illustrated Guide to the Asian Arts of Healing* (Boston, Shambala, 1996).

Francesca Bray, 'Chinese medicine', in W.F. Bynum and Roy Porter (eds.), *Companion Encyclopedia of the History of Medicine* (London, Routledge, 1992), pp. 728–54.

Lawrence I. Conrad, Michael Neve, Vivian Nutton and Roy Porter, *The Western Medical Tradition: 800 BC–1800 AD* (Cambridge, Cambridge University Press, 1995).

Rebecca Flemming, *Medicine and Empire in the Roman World* (Cambridge, Cambridge University Press, 2008).

Mirko D. Grmek, *Diseases in the Ancient Greek World* (Baltimore, Johns Hopkins University Press, 1989).

Ralph Jackson, *Doctors and Disease in the Roman Empire* (London, British Museum, 1988).

Mark Jackson, *Asthma: The Biography* (Oxford, Oxford University Press, 2009).

Carole Reeves, *Egyptian Medicine* (Buckinghamshire, Shire Publications, 1992).

Vivian Nutton, *Ancient Medicine* (London, Routledge, 2004).

Dominik Wujastyk, 'Indian medicine', in W.F. Bynum and Roy Porter (eds.), *Companion Encyclopedia of the History of Medicine* (London, Routledge, 1992), pp. 755–78.

2

Frederick F. Cartwright and Michael Biddiss, *Disease and History* (Stroud, Sutton Publishing, 2000).

Lawrence I. Conrad, 'Arab-Islamic medicine', in W.F. Bynum and Roy Porter (eds.), *Companion Encyclopedia of the History of Medicine* (London, Routledge, 1992), pp. 676–727.

Faye Getz, *Medicine in the English Middle Ages* (Princeton, Princeton University Press, 1998).

John Henderson, *Piety and Charity in Late Medieval Florence* (Oxford, Clarendon Press, 1994).

Peregrine Horden, *Hospitals and Healing from Antiquity to the Later Middle Ages* (Aldershot, Ashgate, 2008).

Kenneth F. Kiple (ed.), *Plague, Pox and Pestilence: Disease in History* (London, Phoenix, 1999).

Michael R. McVaugh, *Medicine before the Plague: Practitioners and their Patients in the Crown of Aragon, 1285–1345* (Cambridge, Cambridge University Press, 1993).

Katharine Park, 'Medicine and society in medieval Europe, 500–1500', in Andrew Wear (ed.), *Medicine in Society: Historical Essays* (Cambridge, Cambridge University Press, 1992), pp. 59–90.

Roy Porter, *Madness: A Brief History* (Oxford, Oxford University Press, 2002).

Carole Rawcliffe, *Medicine and Society in Later Medieval England* (Stroud, Sandpiper Books, 1999).

Carole Rawcliffe, *Leprosy in Medieval England* (Suffolk, Boydell Press, 2006).

Catherine Rider, *Magic and Religion in Medieval England* (London, Reaktion Books, 2012).

Nancy Siraisi, *Medieval and Early Renaissance Medicine* (Chicago, University of Chicago Press, 1990).

Robert Tattersall, *Diabetes: The Biography* (Oxford, Oxford University Press, 2009).

3

Samuel K. Cohn, *Cultures of Plague: Medical Thinking at the End of the Renaissance* (Oxford, Oxford University Press, 2010).

Lawrence I. Conrad, Michael Neve, Vivian Nutton and Roy Porter, *The Western Medical Tradition: 800 BC–1800 AD* (Cambridge, Cambridge University Press, 1995).

Ian Dawson, *Renaissance Medicine* (London, Enchanted Lion Books, 2005).

Peter Elmer (ed.), *The Healing Arts: Health, Disease and Society in Europe, 1500–1800* (Manchester, Manchester University Press, 2004).

Peter Elmer and Ole Peter Grell (eds.), *Health, Disease and Society in Europe, 1500–1800: A Sourcebook* (Manchester, Manchester University Press, 2003).

Roger French, *Dissection and Vivisection in the European Renaissance* (Aldershot, Ashgate, 1999).

David Gentilcore, *Medical Charlatanism in Early Modern Italy* (Oxford, Oxford University Press, 2006).

John Henderson, *The Renaissance Hospital: Healing the Body and Healing the Soul* (Yale, Yale University Press, 2006).

Thomas Wright, *Circulation: William Harvey's Revolutionary Idea* (London, Chatto and Windus, 2012).

4

Andrew Cunningham and Roger French (eds.), *The Medical Enlightenment of the Eighteenth Century* (Cambridge, Cambridge University Press, 1990).

Ole Peter Grell and Andrew Cunningham (eds.), *Medicine and Religion in Enlightenment Europe* (Aldershot, Ashgate, 2007).

Anita Guerrini, *Obesity and Depression in the Enlightenment: The Life and Times of George Cheyne* (Oklahoma, University of Oklahoma Press, 2000).

Joan Lane, *A Social History of Medicine: Health, Healing and Disease in England, 1750–1950* (London, Routledge, 2001).

Roy Porter, 'Was there a medical Enlightenment in England?', *British Journal for Eighteenth-Century Studies*, 5 (1982), pp. 49–63.

Roy Porter, *Health for Sale: Quackery in England 1650–1850* (Manchester, Manchester University Press, 1989).

Roy Porter, *The Enlightenment* (Basingstoke, Macmillan, 1990).

Roy Porter (ed.), *Medicine in the Enlightenment* (Amsterdam, Rodopi, 1995).

Roy Porter, *Bodies Politic: Disease, Death and Doctors in Britain, 1650–1900* (London, Reaktion, 2001).

Dorothy Porter and Roy S. Porter, *Patient's Progress: Doctors and Doctoring in Eighteenth-Century England* (Stanford, Stanford University Press, 1989).

5

Erwin H. Ackerknecht, *Medicine at the Paris Hospital, 1794–1848* (Baltimore, Johns Hopkins Press, 1967).

Roberta Bivins, *Acupuncture, Expertise and Cross-Cultural Medicine* (Basingstoke, Palgrave, 2000).

Deborah Brunton (ed.), *Medicine Transformed: Health, Disease and Society in Europe, 1800–1930* (Manchester, Manchester University Press, 2004).

W.F. Bynum, *Science and the Practice of Medicine in the Nineteenth Century* (Cambridge, Cambridge University Press, 1994).

W.F. Bynum, Anne Hardy, Stephen Jacyna, Christopher Lawrence and E.M. Tansey, *The Western Medical Tradition: 1800–2000* (Cambridge, Cambridge University Press, 2006).

Helen Bynum, *Spitting Blood: The History of Tuberculosis* (Oxford, Oxford University Press, 2012).

Gerald N. Grob, *The Deadly Truth: A History of Disease in America* (Cambridge, Mass., Harvard University Press, 2002).

Christopher Hamlin, *Public Health and Social Justice in the Age of Chadwick: Britain 1800–1850* (Cambridge, Cambridge University Press, 1997).

Anne Hardy, *The Epidemic Streets: Infectious Diseases and the Rise of Preventive Medicine, 1856–1900* (Oxford, Oxford University Press, 1993).

James C. Riley, *Rising Life Expectancy: A Global History* (Cambridge, Cambridge University Press, 2001).

F.B. Smith, *The People's Health, 1830–1910* (London, Croom Helm, 1979).

Simon Szreter, 'The importance of social intervention in Britain's mortality decline c. 1850–1914: a re-interpretation of the role of public health', *Social History of Medicine*, 1 (1988), pp. 1–37.

Nancy Tomes, *The Gospel of Germs: Men, Women, and the Microbe in American Life* (Cambridge, Mass., Harvard University Press, 1998).

Anthony S. Wohl, *Endangered Lives: Public Health in Victorian Britain* (London, Dent, 1983).

Michael Worboys, *Spreading Germs: Disease Theories and Medical Practice in Britain, 1865–1900* (Cambridge, Cambridge University Press, 2000).

6

Virginia Berridge, *Marketing Health: Smoking and the Discourse of Public Health in Britain, 1945–2000* (Oxford, Oxford University Press, 2007).

Roberta Bivins, *Alternative Medicine? A History* (Oxford, Oxford University Press, 2007).

Margaret Brazier and Emma Cave, *Medicine, Patients and the Law* (London, Penguin, 2011).

Roger Cooter, Mark Harrison and Steve Sturdy (eds.), *War, Medicine and Modernity* (Stroud, Sutton Publishing, 1998).

Roger Cooter and John Pickstone (eds.), *Medicine in the Twentieth Century* (Amsterdam, Harwood Academic Publishers, 2000).

Deborah Dwork, *War is Good for Babies and Other Young Children* (London, Tavistock Publications, 1987).

Sander L. Gilman, *Obesity: The Biography* (Oxford, Oxford University Press, 2010).

Anne Hardy, *Health and Medicine in Britain since 1860* (Basingstoke, Palgrave, 2001).

Mark Harrison, *Medicine and Victory: British Military Medicine in the Second World War* (Oxford, Oxford University Press, 2004).

Mark Harrison, *The Medical War: British Military Medicine in the First World War* (Oxford, Oxford University Press, 2010).

Mark Jackson, *Allergy: The History of a Modern Malady* (London, Reaktion, 2006).

Mark Jackson, *The Age of Stress: Science and the Search for Stability* (Oxford, Oxford University Press, 2013).

Christopher Lawrence, *Medicine in the Making of Modern Britain* (London, Routledge, 1994).

Jane Lewis, *The Politics of Motherhood: Child and Maternal Welfare in England, 1900–1939* (London, Croom Helm, 1980).

A. Susan Williams, *Women and Childbirth in the Twentieth Century* (Stroud, Sutton Publishing, 1997).

Index

Page numbers in *italic* denote figures.